NO LONGER PROPERTY OF OLIN LIBRARY WASHINGTON UNIVERSITY

Héctor N. Seuánez

The Phylogeny
of Human Chromosomes

With 49 Figures

Springer-Verlag
Berlin Heidelberg New York 1979

Professor Dr. Héctor N. Seuánez
Department of Genetics, Institute of Biology
Universidade Federal do Rio de Janeiro, Brazil

ISBN 3-540-09303-6 Springer-Verlag Berlin Heidelberg New York
ISBN 0-387-09303-6 Springer-Verlag New York Heidelberg Berlin

Library of Congress Cataloging in Publication Data. Seuánez, Héctor N 1947- The phylogeny of human chromosomes. Includes bibliographies and index. 1. Human chromosomes. 2. Human evolution. I. Title. QH600.S48 599'.9'048732 79-13877

This work is subject to copyright. All rights are reserved, whether the whole or part of the material is concerned, specifically those of translation, reprinting, re-use of illustrations, broadcasting, reproduction by photocopying machine or similar means, and storage in data banks. Under § 54 of the German Copyright Law where copies are made for other than private use, a fee is payable to the publisher, the amount of the fee to be determined by agreement with the publisher.

© by Springer-Verlag Berlin Heidelberg 1979

Printed in Germany

The use of general descriptive names, trade marks, etc. in this publication, even if the former are not especially identified, is not be taken as a sign that such names as understood by the Trade Marks and Merchandise Marks Act, may accordingly be used freely by anyone.

Typesetting, printing, and bookbinding: Brühlsche Universitätsdruckerei, Lahn-Gießen
2131/3130-543210

To María-Helena, María-José, and Pablo

Preface

The question of how man has emerged must be as old as human thought itself. However, it was not until last century that, amidst a storm of opposition and highly emotional criticism, man was first conceived as a product of evolution rather than creation. Moreover, it is not yet thirty years since the chemical composition and molecular structure of the hereditary material was fully understood or the chromosome number of man became known. It should not be surprising then, to find how little, at present, we understand how our genes and chromosomes operate, and how they have evolved during phylogeny.

In this work I have discussed how our own chromosomes have been transmitted and altered as far back as we may trace their phylogeny into the past. To make the work more complete, the composition and evolution of our own genome had also to be consiered in order to understand some of the recent findings at the chromosome level. These have resulted from using methods for localizing repetitive and single copy DNA sequences in chromosomes. Moreover, the development of biochemical methods of studying evolution at the macromolecular level has not only led to a more complete understanding of the evolutionary mechanisms, but has enabled us to make comparisons with evolutionary change at the chromosome level. In addition, a simple reference to the fossil record was necessary, because impressive discoveries in recent years have supplied valuable data on man's evolution.

Before introducing the reader to this work, I wish to express my gratitude to all those who have collaborated with me in many aspects of my research projects and in writing this book. I am very grateful to Dr. David E. Martin of the Yerkes Research Center (Atlanta, USA), Mr. Geoffrey Greed of Bristol Zoo (Bristol, U. K.), and Dr. Robert Martin of Regent's Park Zoo (London, U. K.) for their generous supply of material. I am much indebted to the Director and staff of the MRC Clinical and Population Cytogenetics Unit for their help and collaboration. I am very grateful to Rev. Dr. Alexander King and to Mrs. Magda King for helping me with the manuscript and for their valuable suggestions, and to Dr. Ann Chandley for having read and commented on part of this work. I am also grateful to all those who kindly allowed me to reproduce their illustrations. Finally, I wish to express my gratitude to Dr. Konrad Springer and Dr. Dieter Czeschlik for their most kind assistance and understanding.

Rio de Janeiro, May 1979 　　　　　　　　　　　　　　　H. N. SEUÁNEZ

Contents

Section I. *The Origin of Man* 1

  Chapter 1. *Man, the Most Intelligent Ape* 3
    The Human Paradox 3
    Intelligence, Ape and Man 5
    References 6

  Chapter 2. *The Fossil Record and the Emergence of Modern Man* . . 8
    Africa Versus Asia; Darwin Versus Haeckel 8
    The Fossil Record in Africa 9
    Yet Man Could Have Emerged in Asia 10
    Ramapithecus, Dryopithecus and the Great Apes 11
    From *Homo erectus* to *Homo sapiens* 13
    References 13

  Chapter 3. *Man and His Classification* 15
    The Conflict of Organic and Molecular Evolution 15
    References 19

  Chapter 4. *The Theory of Evolution, Genes, and Chromosomes* . . . 20
    Natural Selection and Mendelian Genetics 20
    Chromosomes, the Vehicles of Inheritance 23
    References 24

Section II. *Cytotaxonomy and the Evolution of Man and the Great Apes* 25

  Chapter 5. *The Chromosomes of Man and the Great Apes. The Inference of Interspecific Homology* 27
    Chromosome Number in the Hominidae 27
    Comparative Studies with Chromosome Banding Techniques . . 28
      1. G- and R-Banding 28
      2. Q-Banding 29
      3. C-Banding 32
      4. G-11 Staining 34
      5. Methylated DNA Sequences 35
      6. T-Banding 35
      7. Ammoniacal Silver (Ag–AS) Staining 36

The Inference of Chromosome Homology Through Different
Degrees of Similarity 36
   1. Chromosomes with Identical Morphology and G- (or R-)
      Banding Pattern in All Species 38
   2. Chromosomes with Very Similar Morphology in All Species   38
   3. Homologous Chromosomes Between Species Which Can Be
      Derived from Each Other by Chromosome Rearrangement
      of G- (or R-) Band Regions 41
   4. Homologous Chromosomes With a Similar Morphology
      but With G- Banding Pattern Which Neither Coincides
      With Nor Can be Derived by Chromosome Rearrangement   41
   5. Chromosomes Having no Similar Counterpart in Any
      Other Species 42
The Y Chromosome 42
References 42

Chapter 6. *Chromosome Heteromorphisms in Man and the Great
Apes as a Source of Chromosome Variation Within Species* 46
Chromosome Heteromorphisms in Man 46
Chromosome Heteromorphisms in the Great Apes 51
   1. Chromosome Heteromorphisms in *Pan troglodytes* . . . 51
   2. Chromosome Heteromorphisms in *Pan paniscus* 52
   3. Chromosome Heteromorphisms in *Gorilla gorilla* 53
   4. Chromosome Heteromorphisms in *Pongo pygmaeus* . . . 55
Phylogenetic Implications of Chromosome Variation in the
Orangutan 58
References 62

Chapter 7. *Chromosome Rearrangement and the Phylogeny of the
Hominidae* 65
Inversions and Telomeric Fusions 65
Implications of Chromosome Rearrangement: a Comparison
with Other Species 69
   1. Inversions 69
   2. Translocations and Centric Fission 70
   3. Telomeric Fusion 70
The Reconstruction of the Ancestral Karyotype of the Hominidae
and the Relationship Between Man and the Great Apes 71
References 77

Chapter 8. *Chromosome Variation Versus Chromosome Fixation* . . 79
    Allopatric and Stasipatric Models of Speciation 79
    References 82

Section III. *Comparative Gene Mapping And Molecular Cytogenetics. A New Approach to Cytotaxonomy* 85

    Chapter 9. *Composition of the Human Genome* 87
        Repetitive and Non-Repetitive DNA Sequences 87
        Palindromes and Tandem Repeats 88
        Satellite DNA and Sequence Heterogeneity 91
        References 93

    Chapter 10. *Evolution of Non-Repetitive DNA Sequences in Man and the Great Apes* 95
        Nucleotide Substitutions and Phyletic Divergence 95
        Man and the Great Apes: Phylogenetic Implications 97
        Is Man an Asian Ape? 98
        References 99

    Chapter 11. *Evolution of Structural Gene Sequences* 101
        Missense Mutations and Amino Acid Substitutions 101
        Molecular Evolutionary Clocks and the Human-Ape Divergence 104
        The Maximum Parsimony Approach and the Decelerated Rates of Molecular Evolution in the Higher Primates and Man . . . 106
        Whence Come Chromosomes? 107
        References 110

    Chapter 12. *Comparative Gene Mapping in Man and Other Primates* 111
        The Evolution of Chromosomes as Syntenic Groups 111
        The Conservation of the Syntenic Groups Among the Hominidae and Cercopithecoidea 115
        Comparative Gene Mapping Between Hominidae–Cercopithecoidea and the Possible Origin of Chromosome 1 in Man . . . 116
        Are Chromosomes Frozen Accidents? 120
        Gene Duplication, Polyploidy, and Evolutionary Frozen Chromosomes 122
        References 126

    Chapter 13. *Evolution of Repetitive DNA Sequences in Man and Other Primates* 128
        Repetitive DNA in the Primates 128
        Repetitive DNA in Man 130

XI

Satellite DNAs in Man and Other Organisms. Possible Explanations of Their Evolutionary Conservation 131
References 135

Chapter 14. *The Chromosome Distribution of Homologous Sequences to the Four Human Satellite DNAs in the Hominidae* 137
The Distribution of Satellite I, II, III and IV in the Human Chromosome Complement 137
The Distribution of Homologous Sequences to the Four Human Satellite I, II, III, and IV DNA in the Chromosome Complement of the Great Apes 139
Interspecific Chromosome Homologies in the Hominidae in Relation to Hybridisation. Independent Amplification of Highly Repetitive DNAs After Speciation 144
References 146

Chapter 15. *DNA Composition of Constitutive Heterochromatin in the Chromosome Complement of Man and the Great Apes* 147
Constitutive Heterochromatin as Demonstrated by C-Banding . 147
G-11 Regions and Satellite III-Rich Regions 153
References 155

Chapter 16. *The Chromosomal Distribution of Ribosomal Genes in Man and the Great Apes* 157
rDNA Genes in Man 157
18S and 28S Cistrons in the Great Apes and Other Primates . . 159
5S rDNA Cistrons in Man and the Great Apes 161
References 162

Chapter 17. *Late DNA Replicating Patterns in the Chromosomes of Man and the Great Apes* 164
DNA Replication at the Chromosome Level 164
DNA Replication Sites in Relation to Chromosome Banding . . 164
The X Chromosome 170
Euchromatin, Heterochromatin and DNA Replication 172
References 176

Chapter 18: *Evolution of Genome Size in Man and the Great Apes* . . 179
The DNA Content of Man and Other Organisms 179
Why Has DNA Content Changed? 181
References 184

Subject Index 187

Section I
The Origin of Man

Chapter 1 Man, the Most Intelligent Ape

The Human Paradox

Man is a relatively new species, for although the first presumed hominids appeared some 15 million years ago, the remains of beings that might properly be considered those of *Homo sapiens,* are not more than 250 000 years old. The rise of man as the most successful of all living species has thus been accomplished in a surprisingly small number of generations. Taking 15 years as the time needed for man to reach puberty, and extrapolating a similar span for our hominid ancestors, the time elapsed since the branching-off of the first hominid from the common hominid-ape stock encompasses 1×10^6 generations. This number may at first appear high; but the same number of generations would cover 250 000 years with a small mammal, e.g. a mouse, which can reproduce at the age of three months. Furthermore, with bacteria capable of dividing under favourable conditions every twenty minutes, they could take as little as 76 years.

But leaving analogies aside, one can go further by more directly examining the very process of change in itself. Since organisms have evolved through having accumulated different kinds of mutations in their genomes along their lines of descent, the explanation of evolution may be discovered in the mechanisms of mutations; how they occur, become fixed, and are transmitted through successive generations. Firstly, it has been demonstrated that in micro-organisms spontaneous mutations occur together with cell division (Vogel et al., 1976) a finding that agrees with the view of the mechanism put forward by Watson and Crick (1953) viz., that spontaneous mutations result from the insertion of a wrong base during DNA replication. Secondly, mutations must necessarily occur in the germ-cell line if they are to be transmitted; otherwise they would inevitably be lost, with no further consequence beyond the life-span of the somatic cells in which they occur. Thus, a better understanding of how much we have evolved in 15 million years can be obtained by estimating the number of cell divisions which our germ-cells have gone through rather than by the number of times entire organisms have been produced. In bacteria, for example, the number of generations equals that of divisions, but in the male mouse it has been estimated that a minimum of 30 cycles of germ-cells divisions takes place per generation (Vogel et al., 1976). In human males it has been estimated that approximately 6×10^8 spermatogonia exist at the age of puberty, a number which may result from approximately 30 dichotomous divisions of a primordial germ-cell; and that at least six more divisions are needed for spermatozoa to be produced. Since germ-cell division occurs continuosly during spermatogenesis in man and mammals, the number of divisions increases with age; nevertheless 36 represents the minimum number of divisions which might occur per generation in human males (Vogel et al., 1976). In females, the highest number of oocytes which is found at five months of pregnancy is approximately 7×10^6, a number that later declines at birth. The number of divisions along the female germ-cell line could then be approximately equal to 21, to which two subsequent divisions must be added for obtaining a mature female gamete independently of the age in which the mature oocyte is produced. Although we do not know if these estimates hold good for extinct hominids, we could envisage that our germ-cell lines have undergone 36×10^6 and 23×10^6 male and female cycles respectively in the last 15 million years, compared to 30×10^6 for male mice in just 250000. Yet, while there is no evidence that in the past 250000 years mice have undergone dramatic change, and neither have bacteria in 76 years, striking changes have occurred in the human lineage. The most remarkable of these has been that affecting the brain size, which was no greater than 450 cc in *Australopithecus africanus* (Holloway, 1972), an ape-like hominid that inhabited the plains of Africa and perhaps also Asia (see Chap. II) as early as 4.5 million years ago. If we consider 1400 cc as the average brain size of modern man, we have increased our brain size by a factor of approximately 3 in only 300000 generations, which roughly comprehends 11×10^6 and 7×10^6 male and female germ-cells division respectively. We can thus imagine that as a very fast-evolving species, we must have accumulated a large number of mutations per germ-cell division and generation to account for the elaborate degree of neural organisation, as well as for many other factors that have allowed us to become the most successful of all living species. We know, however, that natural selection may put a ceiling to the maximum number of mutations that may become fixed in fast-evolving organisms, if these are to survive. Besides, it has been estimated that the rate of structural gene mutation in the high primate lineages has either decelerated (Goodman, 1976) or at most has remained constant (Sarich and Cronin, 1976), to the point that it is not possible to explain organic evolution has being directly derived from simply structural gene mutation (see Chap. 11). Thus, the evolution of man represents the most intriguing of all paradoxes to which no complete answer has been found up to now.

Intelligence, Apes, and Man

For long we have considered ourselves unique in being endowed with an intellect, and for long we have praised our elaborate neural system for all the advantages we enjoy in relation to other species. Man has frequently been defined as the intelligent ape, as the exclusive tool-maker, as the only species capable of communicating through arbitrary symbols (language), and as the only being capable of self-consciousness. Recent discoveries, however, have drastically changed such conceptions up to the point that we have lost the basic criteria of distinctiveness which we thought ourselves to possess. Free-ranging chimpanzees, for example, have been observed to manufacture tools which they are capable of manipulating in a variety of ways to meet the demands of different situations (van Lawick-Goodall, 1968). Moreover, these animals have been observed to engage in planned and deliberate toolmaking, previous to intensive sessions of fishing and hunting (Teleki, 1974). Field studies of many primate species have shown that man is not the only hunter-meat-eating-ape; on the contrary, predatory behaviour has been found to be deliberate in chimpanzees, involving a highly co-ordinated activity of the male members of the group, followed by a behavioural pattern of sharing (Teleki, 1975). Captive chimpanzees have been trained to communicate through gestural signs (Gardner and Gardner, 1971), or with the use of plastic, geometrical symbols that permit the construction of declarative sentences, the understanding of compound sentences, and even the response to interrogatives (Premark, 1971). Furthermore, they have been able to abstract and symbolize objects in their absence, and to describe the symbol of the apple as round, red, and edible. Finally, captive chimpanzees have been able to become aware of their own body, when placed in front of mirrors, by identifying the image with themselves (Gallup et al., 1973), and so indicating that even self-awareness must be dismissed as a unique human psychological trait.

So it is not surprising that, if man's most cherished intellectual attributes are also found in our closest living relatives, the great apes, other no less striking similarities are evident at the organic, biochemical, and chromosome level. King and Wilson (1975) have demonstrated that humans and chimpanzees are so alike at the biochemical level that they resemble each other as do sibling species of other organisms. Comparative studies of the chromosomes of man and the great apes have shown that although differences exist between karyotypes of different species, 99% of the G- or R-band chromosome regions are common to all species, (Dutrillaux, 1975), and it is possible to derive the chromosomes of one species from those of others by chromosome rearrangement (Paris Conference 1971 supplement 1975). One of the implications of such findings is that we must carefully re-evaluate our position in relation to other species, since psychological and biological differences between man and the great apes are a matter of degree rather than a matter of kind. In this respect, we must consider ourselves as the most intelligent of all living beings, rather than the only species endowed with intelligence. Another implication is that a direct comparison between man and his closest living relatives, the great apes, represents a unique insight into our past and a deeper understanding of our origins.

References

Dutrillaux, B.: Sur le nature el l'origine des chromosomes humaines. Paris: L'expansion Scientifique 1975
Gallup, G. G., Jr., McClure, M. K., Hill, S. D., Bundy, R. A.: Capacity for self recognition in differentially reared chimpanzees. Psychol. Rec. *21*, 69–74 (1973)
Gardner, B. J., Gardner, R. A.: Two day communication with an infant chimpanzee. In: Behaviour of non human primates. Schrier, A. M., Stollnitz, F. (eds.), Vol. 4. New York: Academic Press 1971
Goodman, M.: Toward a genealogical description of the primates. In: Molecular anthropology, genes and proteins in the evolutionary ascent of the primates. Goodman, M., Tashian, R. E., Tashian J. (eds.), pp. 321–353. New York, London: Plenum Press 1976
Holloway, R., L., Jr.: Australopithecine endocasts, brain evolution in the Hominoidea, and a model of hominid evolution. In: The functional and evolutionary biology of primates. Tuttle, R. (ed.), pp. 185–203. Chicago: Aldine 1972
King, M. C., Wilson, A. C.: Evolution at two levels in humans and chimpanzees. Science *188*, 107–116 (1975)
Lawick-Goodall, van, J.: The behaviour of free living chimpanzees in the Gombe Stream Reserve. Anim. Behav. Monogr. *1*, 161–311 (1968)
Paris Conference (1971); Supplement 1975. Standardization in human cytogenetics Birth defects: Original article series. XI, 9. New York: National Foundation 1975
Premark, D.: Language in a chimpanzee? Science *172*, 808–822 (1971)
Sarich, V., Cronin, J. E.: Molecular systematics of the primates. In: Molecular anthropology, genes and proteins in the evolutionary ascent of the primates. Goodman, M., Tashian, R. E., Tashian, J. (eds.), pp. 141–170. New York, London: Plenum Press 1976
Teleki, G.: Chimpanzee subsistence technology: Material and skill. J. Hum. Evol. *3*, 375–394 (1974)

Teleki, G.: Primate subsistence patterns: Collector predators and gather-hunters. J. Hum. Evol. *4*, 125–184 (1975)

Vogel, F., Kopun, M., Rathenberg, R.: Mutation and molecular evolution. In: Molecular anthropology, genes and proteins in the evolutionary ascent of the primates. Goodman, M., Tashian, R. E., Tashian, J. (eds.), pp. 13–33. New York, London: Plenum Press 1976

Watson, J. D., Crick, F. H. C.: The structure of DNA. Cold Spring Harbor Symp. Quant. Biol. *18*, 123–131 (1953)

Chapter 2 The Fossil Record and the Emergence of Modern Man

Africa Versus Asia; Darwin Versus Haeckel

When Charles Darwin wrote the "Descent of Man" (1871) there was no evidence that an ape-like hominid ancestor had preceded us on earth. Human fossils had been discovered in Europe since 1848, but these forms, though somewhat primitive-looking, closeley resembled modern man, and were not a proof in favour of Darwin's postulates. Darwin, however, never expected to find the ape-like human ancestor in the European continent, but in Africa. He had clearly recognized that the great apes were man's closest living relatives, and among them he noticed that the African apes, e.g., the chimpanzee *(Pan troglodytes)* and the gorilla *(Gorilla gorilla)* were closer to man than was the orangutan *(Pongo pygmaeus)*, the only great ape of Asian origin. He therefore predicted that man's origin would one day be traced in the tropical forest regions of Africa, where the African great apes now dwell. In the last century, however, there were opinions in favour of the Asian origin of man. Ernst Haeckel in his "Natürliche Schöpfungsgeschichte" (1868) proposed that man could have also originated in South East Asia, an idea which later influenced Dubois to search for fossils in East Java. It was precisely Dubois who in 1891–1894 discovered for the first time the remains of an ape-like hominid which he called "the erect walking ape-man of Java", or *Pithecantropus erectus*. This initial discovery was soon followed by subsequent expeditions to East Java, and by 1909 remains from 11 specimens of *Pithecantropus erectus* had been found there (Jacob, 1973).

These findings led scientists of the time to believe in the Asian origin of man; but in 1925 Dart first challenged this conception with the discovery of a very primitive ape-like fossil form in Taung, which he called the ape-man of South Africa, or *Australopithecus africanus*. However, the hominid status of Dart's *Australopithecus* was questioned, and in the years to come abundant fossil forms continued to appear in Asia. The discovery of a form of *Pithecantropus erectus* in Choukoutien, China, between 1926 and 1937, as well as the large amount of material collected in Central Java between 1931 and 1941, supplied overwhelming evidence in favour of Haeckel's postulates rather than the prediction made by Darwin.

The Fossil Record in Africa

The turn in favour of the African ancestry of man came at the time Broom (1950) discovered new fossil forms in Sterkfontein South Africa. In the years do come similar discoveries in Makapansgat and Swartkrans showed that two different fossil forms had existed. One was the *Australopithecus africanus* described by Dart in 1925, the other correspondend to a more grotesque creature initially called *Paranthropus robustus,* but later considered a more specialized form of Australopithecine, *Australopithecus robustus*. These two specis of *Australopithecus* were clearly more primitive (more ape-like) than *Pithecantropus erectus,* but a comparative study of the pelvic bones of the Australopithecines with those of man and the living species of great apes clearly demonstrated that these creatures were well adapted to a bipedal posture. This finding pointed to the hominid status of *Australopithecus,* in spite of its small brain size which was below 450 cc. Of the two species described, however, *Australopithecus africanus* was the more probable human ancestor, whereas *Australopithecus robustus* represented a branching-off of the original group, which eventually became extinct without undergoing further stages of hominization.

A second stage of discoveries in Africa was that of the Olduvai Gorge. In 1959 L. S. B. Leaky described a hominid fossil, *Zinjanthropus boisei,* which was later considered to be similar to, though not identical with, *Australopithecus robustus,* and was renamed *Australopithecus boisei* (L. S. B. Leaky et al., 1964). An important gap in the lineage leading to man in the African continent was filled when L. S. B. Leaky (1961) found, in Olduvai, remains of a hominid similar to *Pithecantropus erectus.* Thus, similar forms to those found in South-East Asia and China appeared to have existed in Africa, and it was suggested that *Pithecantropus erectus* had evolved from the primitive Australopithecines, although no intermediate form between them has ever been found. It is important to mention at this point, before analysing further discoveries in Africa, that the status of *Pithecantropus erectus* in the hominid lineage was re-examined by Mayr (1963), who pointed out that differences between this fossil hominid and modern man were comparable to differences between species rather than between genera; and he included *Pithecantropus erectus* in the genus

Homo, which thus became *Homo erectus.* A further discovery in Olduvai, however, was to change the taxonomic status of the genus *Homo,* when L.S.B. Leaky et al. (1964) discovered an "advanced" Australopithecine with a cranial capacity above 600 cc, which existed approximately 3 million years ago. This creature, capable of manufacturing primitive tools, represented a significant step forward to hominization, and was considered the earliest species of the genus *Homo,* thus being named *Homo habilis.* This advanced hominid co-existed in Olduvai with the more primitive *Australopithecus boisei* which eventually became extinct. More recently R.E.F. Leaky and Walker (1976) have explored the region of Lake Turkana (formerly Lake Rudolf) and found fossil remains of *Homo erectus* and *Australopithecus boisei* at the same stratigrafic interval, demonstrating that these two hominid lineages also co-existed at the same time in this region of East Africa. The Lower Omo Basin and the Afar depression in East Africa have recently been explored, and a better understanding of Plio/Pleistocene hominids has resulted, mainly by the recovery of well-preserved fossils. One of these skeletons, named "Lucy", has provided a fairly complete picture of a single individual hominid, since over forty specimens belonging to it have been found.

Yet Man Could Have Emerged in Asia

Although the evidence in favour of the African origin of man appears overwhelming, some investigators still believe that man originated in Asia[1]. von Koenigswald (1973) has closely re-examined the relationship between *Australopithecus africanus* and *Australopithecus robustus* based on their dentition, and has concluded that they did not represent different species, but rather members of the same line of evolution; the form described as *Australopithecus robustus* being the late and truncated offshoot of the *Australopithecus* stem. On the other hand, von Koenigswald (1973) considered that a direct ancestor of man was signalised by a pre-*Homo erectus* fossil form found in Java, *Meganthropus paleojavanicus,* whose dentition was somewhat intermediate between *Australopithecus africanus* and *Australopithecus robustus. Meganthropus paleojavanicus* would then represent an "australopithecoid" stage of hominid evolution in Asia, although no cranium of *Meganthropus* has yet been found, and

---
[1] See also Chap. 10.

it is impossible to know whether its brain volume was comparable to that of *Australopithecus*. It must be remarked, however, that one jaw of *Meganthropus* was found to be of the same age as a very early form of *Homo erectus (Pithecanthropus modjokertensis)*, which is approximately 1.9 million years old. This finding proved that in South-East Asia, as in East Africa, more than one hominid lineage co-existed side by side. If *Meganthropus* and some early hominid remains found in China represented an "australopithecoid" stage of hominid evolution, as proposed by von Koenigswald (1973), Australopithecines must then have been widely distributed. This could have resulted from extensive migration of these bipedal ape-like hunters from the place where they originated. One of these places is the Siwalik region in India, where a more primitive hominid, *Ramapithecus*, has been found. *Ramapithecus* seems to be the earliest known hominid which appeared in the Pliocene and anteceded *Australopithecus*. The specimen described by Lewis (1934) consisted only of a maxilla, but its dentition pointed to the fact that it belonged to a hominid, and it probably represents the first creature which branched off from the common stock of the early hominoids into the hominid lineage. Although *Australopithecus* has not been found in the Siwalik region, von Koenigswald (1973) considered that Australopithecus originated in this region which is strategically situated almost equally distant from East Africa and Java, both of them sites where Australopithecine populations could have resulted from migration.

Ramapithecus, Dryopithecus and the Great Apes

Although *Ramapithecus* is apparently the earliest hominid, there is also a controversy over its geographical distribution. Von Koenigswald (1973) considered *Ramapithecus sensu strictu* to have been found only in the Siwalik region of India, whereas Simons and Pilbeam (1965) held that *Ramapithecus* existed in regions as far apart as China, Germany, and Africa. If *Ramapithecus* were already widely spread, it becomes necessary to look even further back into the fossil record to trace the origin of man. Very recently, extensive studies of *Ramapithecus* have been carried out. Its age has been pushed back to about 17 to 7 million years ago, or the Middle Miocene, and even its hominid status has been re-examined (see Kotala, 1977). Simple markers, such as thickness of tooth enamel, are no longer considered discriminative between human and ape lineages; but other factors, such as large size of jaws and cheek teeth in relation to body size, were considered more likely indicators of a hominid status. This criterion forced the inclusion of other miocene fossils e.g., *Sivapithecus* and *Gigantopithecus* into a group with hominid status similar to *Ramapithecus's* (Pilbeam et al., 1977).

Further back in the past the fossil record reveals creatures in which hominid and ape characters blend to indicate a common ancestor of both lineages. Man and the great apes seem to fuse in the Dryopithecines, a group of fossil forms which appeared in the miocene approximately 20 million years ago. The earliest occurrence of *Dryopithecus* has been recorded in East Africa; these creatures have apparently originated from African oligocene hominoids which appeared 40 million years ago. The Dryopithecines migrated from Africa to Asia probably during the Burdigalian-Vindobian stage of the Miocene, a period between 18 and 12 million years ago (Campbell and Bernor, 1976), thus becoming extensively spread. Some species of Dryopithecines have been identified as presumptive ancestors of the chimpanzee *(Pan troglodytes)* and the gorilla *(Gorilla gorilla)*: *Dryopithecus africanus* and *Dryopithecus major,* respectively (Pilbeam, 1969). Within the Burdigalian-Vindobian, the earliest *Ramapithecus* appeared in India and West Pakistan in Asia, and in Fort Ternan in East Africa, so that the possibility exists that man could have originated in both continents. Campbell and Bernor (1976) pointed out that ecological conditions in both continents could have been equally favourable, although the reduction of the forests was more drastic in Asia than in the African sub-Saharan regions, and competition with other grassland-adapted primates was less intense in Asia. The existence of a form of *Homo erectus (Pithecantropus modjokertensis)* 1.9 million years ago suggested that hominids were evolving in Asia earlier in time than in Africa. Moreover, the existence of early hominids in East Africa in the Afar depression, a region geographically close to Asia, may suggest that similar populations once existed in South-West Asia. As a conclusion, the ancestral hominoid stock from which the hominid and the ape lineages gradually diverged originated in Africa and reached Asia by extensive migration. The hominid lineage could equally have evolved in either continent, but the first hominid adaptations seem to have occurred in Asia (Campbell and Bernor, 1976). An intermediate possibility is that man emerged in both continents, with the possibility of intermittent gene exchange between African and Asian populations interrupted by periods of geographical isolation. These periods could have resulted in the divergence of hominid lineages by the branching off of new forms of hominids from the ancestral stock.

From *Homo erectus* to *Homo sapiens*

Although there is general agreement that all human beings belong to the same species and are derived from the same common ancestor, there is no universal agreement on the lineage through which man has emerged. Most palaeontologists and anthropologistsfollow the line *Ramapithecus* → *Australopithecus* → *Homo erectus* → *Homo sapiens*, whereas L. S. B. Leaky (1966) has proposed an alternate lineage substituting *Homo habilis* for *Homo erectus;* this latter being considered as a collateral branch in the hominid lineage. Both hypotheses remain unproven, until intermediate forms of *Homo sapiens* and of either of these presumed ancestors are found. It must be pointed out, however, that the latest populations of *Homo erectus* in China have been estimated as 300,000 years old, and it has been claimed that *Homo erectus* in Australia could have existed as recently as 10,000 years ago (Thorne and Macumber, 1972). The earliest remains of *Homo sapiens* in Swanscombe are estimated as 250,000 years old, so that the temporal distribution of both species of *Homo* practically overlap. A different subspecies of *Homo sapiens (neanderthalensis)* appeared approximately 100,000 years ago and vanished 40,000 years ago, being gradually substituted by modern man *(Homo sapiens sapiens)*. The emergence of modern man as the only surviving subspecies of *Homo sapiens* has given rise to intense controversy between those who believe in a plural ancestry with origins in different geographical locations (polycentric), and others who contend for a single origin in one site, and for later divergence into races through genetic drift operating after human populations were widely dispersed by migration.

References

Broom, R.: The genera and species of the South African fossil-ape men. Am. J. Phys. Antropol. *8*, 1–13 (1950)
Campbell, B. G., Bernor, R. L.: The Origin of the Hominidae: Africa or Asia? J. Hum. Evol. *5*, 441–454 (1976)
Dart, R.: *Australopithecus africanus:* the man ape of South Africa. Nature (London) *115*, 195–199 (1925)
Darwin, C.: The descent of man, and selection in relation to sex. London: John Murray 1871
Dubois, E.: *Pithecanthropus erectus,* eine menschenähnliche Übergangsform aus Java. Batavia: Laudesruckerei 1894
Haeckel, E.: Natürliche Schöpfungsgeschichte. Berlin: Reimer 1868
Jacob, T.: Palaeoanthropological discoveries in Indonesia with special reference to the finds of the last two decades. J. Hum. Evol. *2*, 473–485 (1973)
Koenigswald, von, G. H. R.: *Australopithecus, Meganthropus* and *Ramapithecus.* J. Hum. Evol. *2*, 487–491 (1973)

Kotala, G. B.: Human evolution: Hominoids of the miocene. Science *197*, 244–245 (1977)
Leaky, L. S. B.: A new fossil skull from Olduvai. Nature (London) *184*, 491–493 (1959)
Leaky, L. S. B.: New finds at Olduvai Gorge. Nature (London) *189*, 649–650 (1961)
Leaky, L. S. B.: *Homo habilis, Homo erectus* and the australopithecines. Nature (London) *209*, 1279–1281 (1966)
Leaky, L. S. B., Tobias, P. V., Napier, J. R.: A new species of the genus *Homo* from Olduvai Gorge. Nature (London) *202*, 7–9 (1964)
Leaky, R. E. F., Walker, A. C.: *Australopithecus, Homo erectus* and the single species hypothesis. Nature (London) *261*, 572–574 (1976)
Lewis, G. E.: Preliminary notice of new man-like apes from India. Scientific results of the Yale North-India expedition. Report on vertebrate palaeoantology. Am. J. Sci. *27*, 163–179 (1934)
Mayr, E.: The taxonomic evolution of fossil Hominids. In: Classification and human evolution. S.L. Washburn (ed.). Viking Fund Publications in Anthropology. *37*, 1963
Pilbeam, D. R.: Tertiary pongidae of East Africa. New Haven Connecticut: Peabody Museum of Natural History and Department of Anthropology. Yale Univ. Bull. *31* (1969)
Pilbeam, D., Meyer, G. E., Badgley, C., Rose, M. D., Pickford, M. H. L., Behrensmeyer, A. K., Ibrahim Shah, S. M.: New hominoid primates from the Siwaliks of Pakistan and their bearing on hominoid evolution. Nature (London) *270*, 689–695 (1977)
Simons, E. L., Pilbeam, D. R.: Preliminary division of the Dryopithecinae (Pongidae, Anthropoidea). Folia Primatol. *3*, 81–152 (1965)
Thorne, A. G., Macumber, P. G.: Discoveries of late pleistocene man at Kow Swamp, Australia. Nature (London) *238*, 316–319 (1972)

Chapter 3 Man and His Classification

The Conflict of Organic and Molecular Evolution

Classically, taxonomists relied on anatomical characteristics of living beings or of extinct fossils for assigning organisms to taxons, so that the classification indicated a hierarchical order of macro-structural complexity. For pre-evolutionary taxonomists, the main task of systematics was (1) to classify, (2) to name, and (3) to indicate degrees of resemblance between organisms. After the theory of evolution was put forward, systematics rapidly became on evolutionary science whose most important aim was to show relationships by descent. Simpson (1944) championed this latter approach to systematics, and moreover, used this classification to estimate the rate of structural evolution of organisms during phylogeny. A good example provided by the lineage of the horse (genus *Equus*) which has undergone eight successive genera in roughly 45 million years. This works out a approximateely 0.18 genera per 10^6 years, a standard evolutionary rate of organic change which Simpson called "horotelic". There was also a second approach to the study of organic evolution, based on quantitative characteristics, e.g., brain size, which, as we have previously stated, has increased in the human lineage by a factor of approximately 3 in about 5 million years.

On the other hand, evolution has also been operative at the molecular level, which accounts for changes in the immense number of macromolecular components of living beings since life first appeared on earth. The development of powerful biochemical techniques in analysing protein structure has enabled us to compare similar proteins of different species and estimate the number of aminoacid substitutions which have occurred between them. One method used relies on the property of proteins of being antigenically different, due to a change in primary structure. In consequence, "immunological distances" can be estimated between species by tests of microcomplement fixation. Goodman (1975) has used this method to study a large variety of species, and construct phylogenetic trees by a cladistic analysis, on the assumption that differences between species can be best explained by the minimum number of amino acid substitutions between them, or parsimonious evolution. Another method of analysing proteins is to use electrophoretic migration that detects differences between proteins whenever amino acid substitutions affect their total charge. King and Wilson (1975) have estimated that approximately $1/3$ of point mutations are detectable by this method, and results obtained by electrophoresis coincide well with those obtained by microcomplement fixation between human and chimpanzee proteins. A third method is to compare proteins in their respective amino acid sequences (see Dayhoff, 1972), by which complete information on the number and the kind of amino acid substitutions can be obtained. However, it must be pointed out that studies at the protein level offer only a limited insight into the differences in kind and degree of macromolecular evolution which species have undergone. This is because, firstly, protein analysis underestimates the number of nucleotide substitutions at structural cistrons due to the degeneracy of the genetic code, for redundancy makes most third-position nucleotide replacements meaningless in terms of amino acid substitutions. Secondly, structural genes must comprise only a very small percentage of the overall eukaryote genome. If it were not so, the extrapolated deleterious mutation rate in the number of theoretically functional loci, (10^{-5} per locus per generation), as observed in man and mammals, would become approximately equal to 30 (Ohno, 1972), thus representing a genetic load that no organism could bear. The finding that substantial amounts of the eukaryote genome were composed of apparently functionless repetitive DNA sequences (Chap. 9) is good evidence that the structural genes present in few or single copies per haploid genome must represent but a small fraction of the total DNA. For these reasons, a comparison of species at the DNA level may lead to a more accurate assessment of the change(s) that have taken place, even though it is more difficult to make individual comparisons of purified single-copy DNA sequences, as one can do with individual proteins.

In spite of the above-mentioned limitations, molecular evolution has supplied most valuable information on how genes have evolved. Its study has shown, for example, that man and the great apes form a cluster of very similar species whose mutual resemblance is closer than is the resemblance of either to the gibbons (Goodman, 1975). A clear example of this resemblance is supplied by the results obtained by King and Wilson (1975) showing that the average human and chimpanzee protein differs in only 7.2–8.2 amino acid sites per 1000 substitutions, actually a difference less than 1% between the two species at the protein level. Kohne et al. (1972) have estimated that these two species have undergone approximately 2.4% nucleotide replacements in their non-repetitive DNA fraction since their separation (Chap. 10), a smaller difference than that between man and the gibbon, or man and monkeys. Results of this kind are sharply in contrast with previous observations at the macrostructural level, in which man was clearly considered as the most radically distinct of all hominoids, whereas the chimpanzee, for example, retained many of the characteristics of the ancestral form from which it had derived (Simpson, 1963). It was mainly these differences, so evident at the anatomical level, which led Simpson (1945) to assign man to a distinct family (Hominidae) of which he was the exclusive member among living species, whereas the great apes and the gibbons were assigned to a different family, the Pongidae (see Table 3.1). However, the evidence supplied by molecular evolution led Goodman (1975) to propose that man and the great apes should be included in one family, the Hominidae, whereas the gibbons and siamangs should be grouped into a different family, the Hylobatidae (see Table 3.2).

Table 3.1. Classification of Simpson (1945)

Superfamily Hominoidea (Simpson, 1931)
  Family Pongidae (Elliot, 1913)
    Subfamily Ponginae (Allen, 1925)
      Pongo (Lacepede, 1799)
        pygmaeus (Linnaeus, 1760)   (Orangutan)
          pygmaeus (Linnaeus, 1760)   (Borneo orangutan)
          abelii (Lesson, 1826)   (Sumatra orangutan)
      Pan (Oken, 1816)   (Chimpanzee)
        paniscus (Schwarz, 1929)   (Pygmy chimpanzee)
        troglodytes (Blumenbach, 1799)   (Chimpanzee)
      Gorilla (I. Geoffroy, 1852)
        gorilla (Savage & Wyman, 1847)
          gorilla (Savage & Wyman, 1847)   (Coast gorilla)
          beringei (Matschei, 1903)   (Mountain gorilla)
    Subfamily Hylobatinae (Gill, 1872)
      Hylobates (Illiger, 1811)   (Gibbon)
        agilis (Cuvier, 1821)   (Dark-handed gibbon)
        concolor (Harlan, 1826)   (Black gibbon)
        hoolock (Harlan, 1834)
        klossii (Miller, 1903)   (Kloss's gibbon)
        lar (Linnaeus, 1771)
          lar (Linnaeus, 1771)   (White-handed gibbon)
          pileatus (Gray, 1861)
        moloch (Audebert, 1797)   (Sunda island gibbon)
        mulleri (Martin, 1841)
      Symphalangus (Gloger, 1841)   (Siamang)
        syndactylus (Raffles, 1821)
  Family Hominidae (Gray, 1825)
    Homo (Linnaeus, 1758)
      sapiens (Linnaeus, 1758)   (Man)

Table 3.2. Classification of the Hominoidea following Goodman (1975)

Superfamily Hominoidea	
Family Hylobatidae	
Subfamily Hylobatinae	
Hylobates	(Gibbons)
Symphalangus	(Siamangs)
Family Hominidae	
Subfamily Ponginae	
Pongo	
pygmaeus	(Orangutan)
Subfamily Homininae	
Pan	
troglodytes	(Chimpanzee)
paniscus	(Pygmy chimpanzee)
Gorilla	
gorilla	(Gorilla)
Homo	
sapiens	(Modern man)

Since man's position in the classification will ultimately depend on our ability to integrate the information supplied by comparative studies between man and other non-human primates, as well as the information from the fossil record, it is necessary to evaluate man's position from a cytotaxonomic point of view, a point that will be more fully dealt with in the following chapters. In this respect, man and the great apes form a cluster of species in which the chromosome number has been maintained within narrow change (46 in man vs. 48 in the great apes), whereas in the next closely related primates, the gibbons and siamangs, the degree of variation in chromosome number ranges from 44 to 52. However, what makes man and the great apes an extremely close group from a cytotaxonomic point of view is not only number, but the fact that chromosome morphology and banding patterns are very similar between all species, whereas very few human (or great ape) chromosomes can be identified as similar to those of gibbons or siamangs. As will be discussed in Chap. 12, several similar chromosomes can be identified between man and the great apes on one side and some Cercopithecoid species on the other, but, nevertheless, man and the great apes still resemble each other more closely than either of them resembles a Cercopithecoid monkey at the chromosome level, both in diploid number or morphology. Even though syntenic homologies have been found between some chromosomes of Cercopithecoid monkeys and human (and great ape) chromosomes, these latter show some distinctive characteristics. One of them, for example is the presence of sequences homologous to human satellite

DNAs, which have been detected and localized in the chromosome complement of the great apes, but not of any other primate species (Chap. 14). Thus, the integration of man and the great apes into a single family advanced by Goodman (1975) at the protein level can also be established at the chromosome level. Moreover, the proposition that the orangutan (Ponginae) is further away both from man and the African apes (Homininae) at the molecular level is supported by some good evidence also at the chromosome level: brilliant quinacrine fluorescence is restricted to man and the African apes, while it is absent in the orangutan. The probable reason is that this kind of chromatin appeared in the common ancestor of man and the African great apes after the branching off of the orangutan from the common stock (Pearson, 1973). Since further evidence of this remarkable similarity between human and great ape chromosomes will be extensively analysed in the chapters to come, it is unnecessary to describe these points in detail at this stage. However, for all the above mentioned reasons the classification proposed by Goodman (1975), is preferable, and so the term "Hominidae" used in the following chapters will designate a family which includes man and the great apes as different species, contrary to the previous classification (Simpson, 1945) in which man was considered as the unique living species of this family.

References

Dayhoff, M.O. (ed.): Atlas of protein sequences and structure; National Biochemical Research Foundation. Vol. 5. Washington D.C.: Georgetown University Medical Center 1972

Goodman, M.: Protein sequence and immunological specificity. Their role in phylogenetic studies of the primates. In: Phylogeny of the primates. Luckett, W.P., Szalay, J.S. (eds.), pp. 219–248. New York: Plenum Press 1975

King, M.C., Wilson, A.S.: Evolution at two levels in human and chimpanzees. Science *188*, 107–116 (1975)

Kohne, D.E., Chiscon, J.A., Hoyer, B.H.: Evolution of primate DNA sequences. J. Hum. Evol. *1*, 627–644 (1972)

Ohno, S.: An argument for the genetic simplicity of man and other mammals. J. Hum. Evol. *1*, 651–662 (1972)

Pearson, P.L.: The uniqueness of the human karyotype. In: Chromosome identification techniques and applications in biology and medicine. Caspersson, T., Zech, L. (eds.), pp. 145–151. New York, London: Academic Press 1973

Simpson, G.G.: Tempo and mode in evolution. New York: Columbia University Press 1944

Simpson, G.G.: The principles of classification and classification of the mammals. Bull. Am. Mus. Nat. Hist. *85*, 1–350 (1945)

Simpson, G.G.: In: Classification and human evolution. Washburn, S.L. (ed.). Chicago: Aldine 1963

Chapter 4 The Theory of Evolution, Genes, and Chromosomes

Natural Selection and Mendelian Genetics

The idea that all living beings including man had originated from pre-existing forms of life rather than by the acts of divine creation was one of the most revolutionary conceptions to challenge the established scientific thought of any time. Although in the early 19th century Lamarck had already questioned the idea of creation, it was not until Charles Darwin that a coherent theory of evolution was first presented. In "The Origin of Species" (1859), Darwin postulated that evolution had taken place through natural selection, or the survival of the fittest. He pointed out, moreover, that evolution could never have occurred, had not the fittest been at the same time also fertile; otherwise, their predominance in the struggle for life would have been transitory, with no further consequence beyond their individual life-spans. As Darwin correctly observed, intraspecific variation was necessary for natural selection to act upon: it allowed the organisms better adapted to their environment to survive, while is condemned the less fit to perish.

On the other hand, the mechanisms by which living beings became endowed with the different characters that would determine either their survival or extinction were a mystery to Darwin as well as to the scientists of his time, who had no clear understanding of genetics. The correct explanation of how this occurred came from the work of Gregor Mendel, who in 1865 discovered the basic mechanisms by which hereditary factors are transmitted. His findings showed that organisms with sexual reproduction are capable of producing genetically heterogeneous offspring, resulting from the segregation and independent assortment of their genes into gametes. In this way, different gene combinations can be produced in proportions predictable by simple probability estimates. His findings also showed that the physical attributes of organisms, or phenotypes, depended on their genetic constitutions, or genotypes. Intraspecific variation could thus be explained by the immense variety of possible gene combinations resulting from heterozygosity at a large number of loci. Natural selection was therefore conceived as the predominance of those phenotypes which advantageous gene combinations enabled to compete successfully for survival in their environment. This implied that the incidence of genes in a population fluctuated according to the direction and intensity of selection, those conferring successful adaptations becoming more frequent, until they reached ultimate fixation. However, if genes thus became fixed by natural selection, it follows that a price had to be paid for each locus subject to selection, because a considerable number of individuals that lacked the favourable gene would either be eliminated in each generation, or would exhibit lower fertility.

In 1957 Haldane estimated that for a diploid population with moderately selective advantage, the "cost" required for substituting one allele for another would equal 10 to 100 times its population size. Moreover, in populations subject to so intensive a selection that 50% of their members were eliminated per generation, Haldane calculated that one gene substitution could take place in as little as 43 generations. Since, however, he considered that natural selection would normally operate under less drastic conditions, he estimated that the rate of gene substitution would be approximately one per 300 generations. Two important conclusions follow from Haldane's approach: (1) that periods of fast adaptations require high selective costs in terms of the number of individuals that perish, and (2) that the number of gene loci involved in such adaptations must be very few, if species are not to exterminate themselves through natural selection. Since Haldane's estimates have been frequently quoted, it is important to mention that they are based on the assumption of "multiplicative fitness", as against the "threshold fitness" discussed by Maynard Smith (1968).

Multiplicative fitness implies that selection acts independently at all loci. Thus, in an haploid organism, for example, if we assume that *A* shows a relative fitness=1 compared to $a=(1-k)$, and *B* also shows a relative fitness=1 compared to $b=(1-l)$, only *AB* organisms would survive, because selection would operate on *A* versus *a*, regardless of *B*, and on *B* versus *b*, regardless of *A*; the relative fitness of *AB, aB, Ab* and *ab* being 1: $(1-k)$: $(1-l)$: $(1-k)(1-l)$. Maynard Smith (1968) commented that Haldane's assumptions, when applied to all gene loci, overestimate the cost of natural selection, whereas the assumption that selection operates on a "threshold model" allows a number of gene loci to be selected that exceeds Haldane's estimate by several orders of magnitude. If, for example, selective deaths were confined to organisms which lack all favorable alleles *(ab)*, those which lack only one or another, i.e., the *Ab* and *aB* individuals, would also survive together with the *AB* individuals. Organisms would thus be eliminated by selection only when lacking a minimum number of favourable alleles, or in other words, only when falling below a "threshold" of fitness.

Estimates based on the latter assumption show drastically different results from Haldane's. Maynard Smith (1968) found, for example, that selection could act simultaneously at 51,000 loci with a selective advantage of 1% in a diploid organism undergoing 50% of selective deaths per generation if the frequency of the less favourable allele equalled 0.5. However, under similar conditions, but assuming multiplicative fitness, the number of loci at which selection could operate would not exceed 138. This discrepancy, we must point out, is the result of extreme estimates of the number of loci that may be subject to selection, since fitness is neither universally multiplicative, nor can it always be explained by a threshold model. It must be stressed, however, that Haldane's estimate of 300 generations being needed to accomplish one gene substitution, according to Maynard Smith, grossly underestimates the rate of evolution. It should not lead to the inference that the large proportion of gene substitutions which have taken place during evolution could not have been the result of selective fixation, as proposed by Kimura (1968).

A further important fact to be noted is that neither natural selection nor Mendel's principles could explain how new genes were formed by organisms, a point that was crucial for understanding evolution. It soon became clear that natural selection, per se, could never have given rise to speciation, had not new genes been produced by mutation. Moreover, as Ohno pointed out in 1970, a substantial gene duplication must also have played an essential part in evolution, being a necessary condition for accumulating mutations in the genome free from selective pressures.

Finally, if speciation took place by the fixation of successful gene combinations, either as a result of natural selection or drift, reproductive barriers between incipient species must have been created at some stage during phylogeny, since no significant leap forward in evolution could have occurred without genetic isolation. Species have diverged and maintained their difference from each other ever since, as a consequence of sharing their genetic pool exclusively among their own members; otherwise, if gene flow were established between two species, the two would ultimately blend into one. We could therefore define species in terms of gene exchange rather than by structural organization, although the definition must be restricted to organisms with sexual reproduction. Organisms that reproduce asexually have no means of exchanging genes with other members of the same species, and except for mutation, are doomed to produce genetically identical offspring. Since, however, the above-mentioned definition applies to the vast majority of organisms that have succeeded in adapting to their environments, we can aptly conceive a species as being a population of individuals within which gene exchange naturally occurs, and which maintains its uniqueness by the fact that it constitutes a genetic isolate, its gene pool showing a discrete distribution and no overlapping with that of other species.

Chromosomes, the Vehicles of Inheritance

Early in this century, Morgan (1911) established that genes were linearly aligned on visible nuclear structures, or chromosomes, a discovery that was to modify substantially Mendel's postulates. Whereas Mendel's principle of dominance and recessiveness was still applicable to genes, segregation and independent assortment were found to apply more precisely to the behaviour of chromosomes. The mode of transmission of groups of genes inside chromosomes was then explained by linkage, and the less frequent production of recombinants by crossing over. The study of the chromosome complement of different species and of the meiotic cycles of many organisms led to two important discoveries: one, that each species had a distinct chromosome constitution (chromosome number and morphology), and two, that a normal chromosome constitution was necessary for meiosis to occur undisturbed in an individual with sexual reproduction.

If man, then, has emerged from an extinguished miocene ancestor from which our closest living relatives, the great apes, have also evolved, what would this imply? It would imply that an original population of breeding individuals with presumably the same chromosome constitution branched off into different subpopulations which later became reproductively isolated, thus splitting the original genetic pool of our common ancestor into separate genetic enclaves. Although we cannot prove whether chromosome change was a cause or a consequence of speciation, there is a possibility that chromosome rearrangement might have played a significant role in the establishment of permanent reproductive barriers beteeen each subpopulation of emerging hominids and primitive apes. Some of these intriguing mechanisms can be envisaged if the chromosomes of our closest living relatives, the great apes, are compared to the human chromosome complement. Comparative studies have shown that chromosomes of one species can be derived from those of another species by assuming that chromosome rearrangement, such as inversion or fusion, took place between them at some stage during phylogeny. The availability of chromosome-banding techniques has made such comparisons possible in great detail. Studies of comparative gene mapping between man, the great apes, and some species of old world monkeys have clearly demonstrated interspecific homologies between syntenic groups in general coincidence with morphological observations. Thus, in spite of morphological change due to chromosome rearrangement, the phylogeny of our chromosomes can be traced back as far as 35 million years in the catarrhine lineage, a period during which there has been a remarkably evolutionary conservation at the chromosome level, in spite of the accelerated rate of organic evolution. The comparison of human and non-human primate chromosomes in relation to the patterns of late DNA-synthesis has also demonstrated that these have remained basically unmodified for a period that goes back at least to the time when man and the great apes diverged from each other. The possibility of detecting and localizing highly repetitive DNA sequences in chromosomes by techniques of in situ hybridisation has demonstrated that some of our repetitive DNAs are also present in the genome of the great apes, although their distribution in chromosomes might not always be strictly coincident.

Since the evolution of man is ultimately the history of how our genes have been transmitted and diverged ever since we branched off from the common stock with the great apes, a substantial part of this process can be understood by studying our chromosomes, and tracing back into the past the presumptive stages which they have gone through. It is with this hope that I introduce the reader to the following chapters.

References

Darwin, C.: The origin of species by means of natural selection, or the preservation of favoured races in the struggle for life. London: John Murray 1859

Haldane, J. B. S.: The cost of natural selection. J. Genet. 55, 511–524 (1957)

Kimura, M.: Evolutionary rate at the molecular level. Nature (London) 217, 624–626 (1968)

Maynard-Smith, J.: Haldane's Dilemma and the rate of evolution. Nature (London) 219, 1114–1116 (1968)

Mendel, G.: Versuche über Pflanzen-Hybriden. Verh. Naturforsch. Ver. Brünn 4, 3–47 (1865)

Morgan, T. H.: An attempt to analyse the constitution of the chromosomes on the basis of sex-limited inheritance in *Drosophila*. J. Exp. Zool. 11, 365–413 (1911)

Ohno, S.: Evolution by gene duplication. Berlin, Heidelberg, New York: Springer 1970

Section II
Cytotaxonomy and the Evolution of Man and the Great Apes

Chapter 5   The Chromosomes of Man and the Great Apes. The Inference of Interspecific Homology

Chromosome Number in the Hominidae

Since the time when Tjio and Levan (1956) and Ford and Hamerton (1956) first reported that the normal chromosome number of man was 46, thus ending a long-standing controversy, substantial improvements in the methods of obtaining chromosome preparations have been made. These have permitted the study of other species, among which are included man's closest living relatives, the great apes: *Pan troglodytes, Pan paniscus, Gorilla gorilla* and *Pongo pygmaeus*. Except for an early report of Yeager et al. (1940), who had observed meiotic divisions in preparations of the testis of a chimpanzee *(Pan troglodytes)*, nothing was known about chromosome number in the great apes. Based on the number of observed bivalents, Yeager et al. (1940) had correctly reported the diploid chromosome number of this species to be 2n=48. Twenty years later, Young et al. (1960) confirmed these findings in the chimpanzee, using chromosome preparations obtained from bone marrow. Later on, the chromosomes of the other species of great ape were studied for the first time (Chiarelli, 1961; Hamerton et al., 1961; Chiarelli, 1962), and their diploid chromosome number was also found to be 48. The karyotypes of the great apes were compared to that of man as a first attempt to study the phylogeny of human chromosomes (Chu and Bender, 1962; Bender and Chu, 1963; Hamerton et al., 1963; Klinger et al., 1963; Egozcue, 1969). An interesting comparison between the chromosomes of *Pan troglodytes* and man was made by McClure et al. (1969, 1971) who first reported a chromosomal numerical aberration in a non-human primate. The propositus was a new-born chimpanzee that was trisomic for a small acrocentric chromosome (No. 22 in their classification). The clinical condition in the propositus resembled that of Down's syndrome in man, and the trisomic chromosome was recognized as identical to chromosome 21 in man. It was thus evident that a similar clinical condition was produced in two phylogenetically related species owing to an identical kind of chromosomal aneuploidy. The diagnosis of the trisomy and the identification of the extra chromosome was later confirmed with Q-banding (Benirschke et al., 1974).

Comparative Studies with Chromosome Banding Techniques

The development of chromosome banding techniques in the early seventies stimulated a number of workers to make detailed studies of the chromosomes of the great apes and to compare their banding patterns with those of the human chromosome complement (Borgaonkar et al., 1971; Pearson et al., 1971; Chiarelli and Lin, 1972; de Grouchy et al., 1972 and 1973; Turleau et al., 1972 and 1975; Bobrow and Madan, 1973; Dutrillaux et al., 1973; Egozcue et al., 1973a, b; Khudr et al., 1973; Lejeune et al., 1973; Lin et al., 1973; Pearson, 1973; Turleau and de Grouchy, 1973; Warburton et al., 1973; D.A. Miller et al., 1974; Bobrow, 1975; Dutrillaux, 1975 et al., 1975b, c; Seuánez et al., 1976a, b; Seth et al., 1976; Bogart and Benirschke, 1977a, b).

1. G- and R-Banding

With G-banding techniques, (Gallimore and Richardson, 1973), and with R-banding techniques, (Dutrillaux and Lejeune, 1971), it was possible to make detailed comparisons between human and great ape chromosomes and to identify presumed homologues between different species based on their degree of similarity. As an example, chromosome pair 6 in man was so similar to a chromosome pair of all the other species in size, arm ratio, and G- or R-banding, that it was considered to have been conserved unchanged from the common ancestor of man and the great apes. However, in other cases, interspecific homology was less evident, and this situation led investigators to make assignments following different criteria. Fortunately, a standardized criterion of nomenclature and presumptive homology between human and great ape chromosomes was established in the report of the Paris Conference (1971) supplement (1975), and these recommendations have been followed in this work. Each chromosome is designated by a combination of three letters, or acronym, denoting the name of the species to which it belongs followed by its number; the proposed acronyms being: HSA for *Homo sapiens;* PTR for *Pan troglodytes;* PPA for *Pan paniscus;* GGO for *Gorilla gorilla* and PPY for *Pongo pygmaeus.* The standardized G-band karyotypes of the four species of great ape are shown in Figs. 5.1–5.4, and the human karyotype is shown in Fig. 5.5 as a reference to which chromosomes of other species can be compared. Looking at these figures it can be appreciated, even by those who are not cytogeneticists, that some chromo-

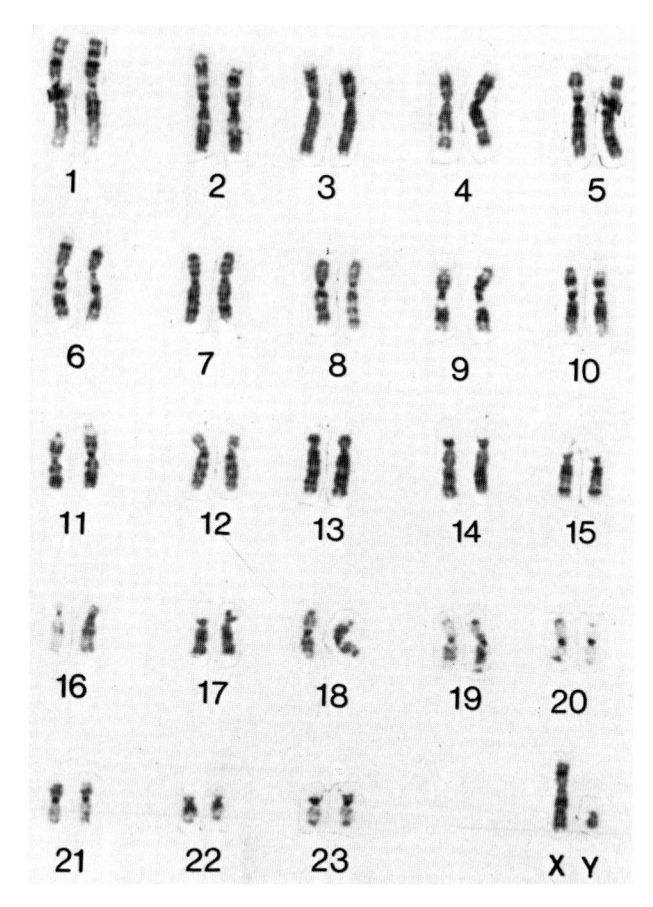

Fig. 5.1. G-band karyotype of *Pan troglodytes*

somes are practically identical when comparisons are made between any pair of species.

2. Q-Banding

The introduction of fluorescent dyes to stain chromosomes by Caspersson et al. in 1968 allowed the identification of homologous chromosomes by their banding patterns, revealed by the differential intensity of staining along the chromatids. With quinacrine dihydrochloride, the observed variation in fluorescence intensity of the human chromosome complement can be described as either brilliant, intense, medium, pale, or negative (Paris Conference, 1971). Brilliant fluorescence has also been found in

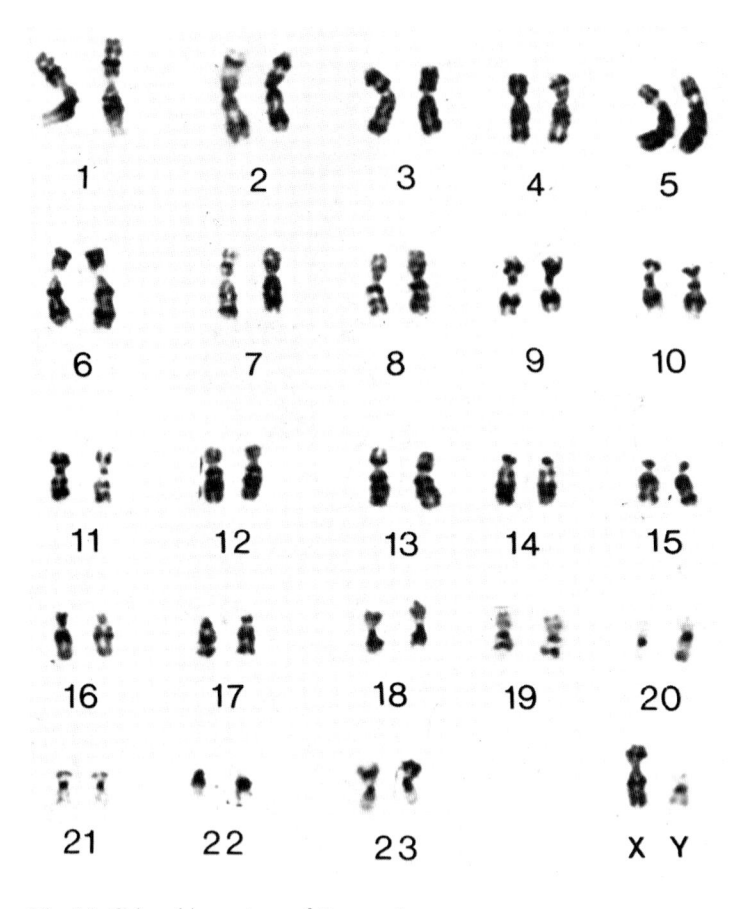

Fig. 5.2. G-band karyotype of *Pan paniscus*

the chromosome complement of the African apes, the two species of chimpanzee and the gorilla, but it is absent in the orangutan and the gibbon. Moreover, man and the African apes were found to be the only species showing brilliant fluorescence among mammals (Pearson et al., 1971); man and the gorilla being the only ones showing a brilliant fluorescent distal region at the Y chromosome long arm. As illustrated in Fig. 5.6, brilliant fluorescence can be found in many chromosomes that are recognized as homologues between species, and moreover, at similar regions between interspecific homologues. For example, chromosome 21 in man and its homologues in the African apes (PTR 22, PPA 22, and GGO 22) may show brilliant fluorescence in their short-arm region. Findings of this kind led Pearson (1973) to propose that brilliant fluorescence has appeared in a common ancestor of man and the African apes (or

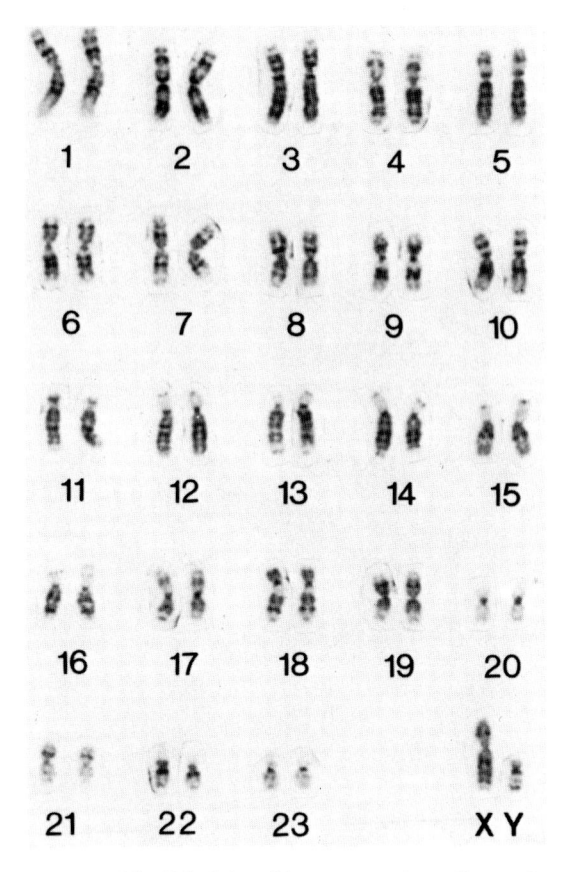

Fig. 5.3. G-band karyotype of *Gorilla gorilla*

the subfamily Homininae according to Goodman, 1975) after the orangutan (Ponginae) had branched off from the common stock (the Hominidae).

In man, brilliant fluorescent regions may be found at the distal region of the Y chromosome long arm, the short-arm-satellited region of the acrocentric chromosomes (pairs 13, 14, 15, 21, and 22) and the centromeric region of chromosomes 3 and 4. In the two species of chimpanzee, brilliant fluorescent regions are restricted to some of the autosomes (pairs 14, 15, 17, 22, and 23), but their Y chromosome is pale staining. Chromosome 23 in *Pan troglodytes* is, however, an acrocentric chromosome, whereas in *Pan paniscus* it is a small metacentric; nevertheless both may show brilliant fluorescent regions (see Fig. 6.5). In the gorilla, brilliant fluorescent regions may be found in both the Y chromosome, and in some autosomes (pairs 3, 12, 13, 14, 15, 16, 22, and 23).

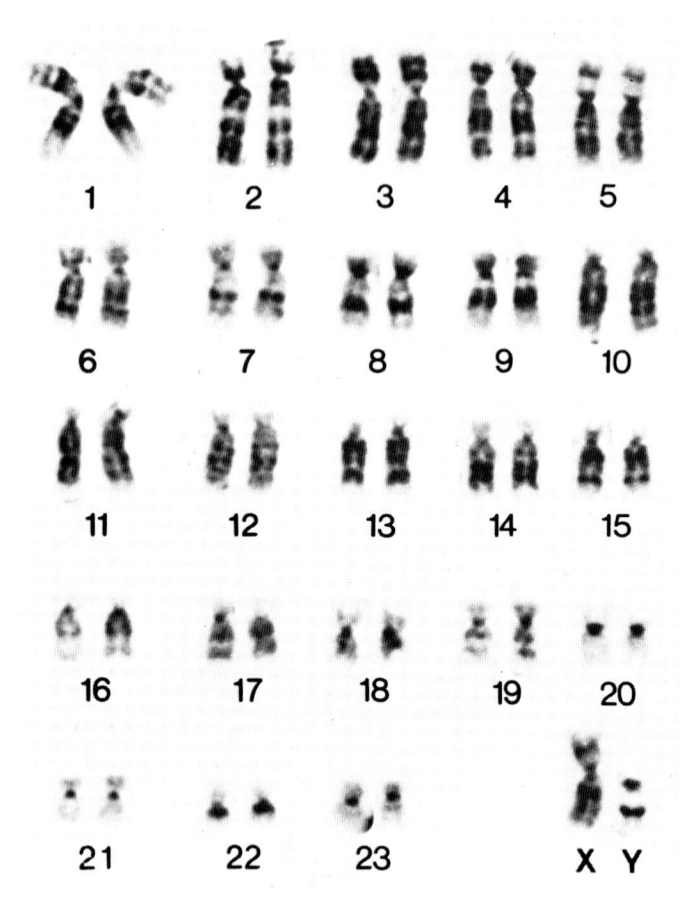

Fig. 5.4. G-band karyotype of *Pongo pygmaeus* (Sumatran specimen)

Terminal Q-bands of medium intensity can be observed at the telomere of many chromosome arms in the two species of chimpanzee and the gorilla, but not in man or the orangutan (see Fig. 6.6). These regions are positively C-banded and are probably genetically inert.

3. C-Banding

With the C-banding technique as a method of demonstrating constitutive heterochromatin (Arrighi and Hsu, 1971) two important distinctions can be made. One is that some species show terminal C-band regions, e.g., chimpanzee and the gorilla, while others do not, e.g., man and the orangutan. Another is that some species show a positively C-banded

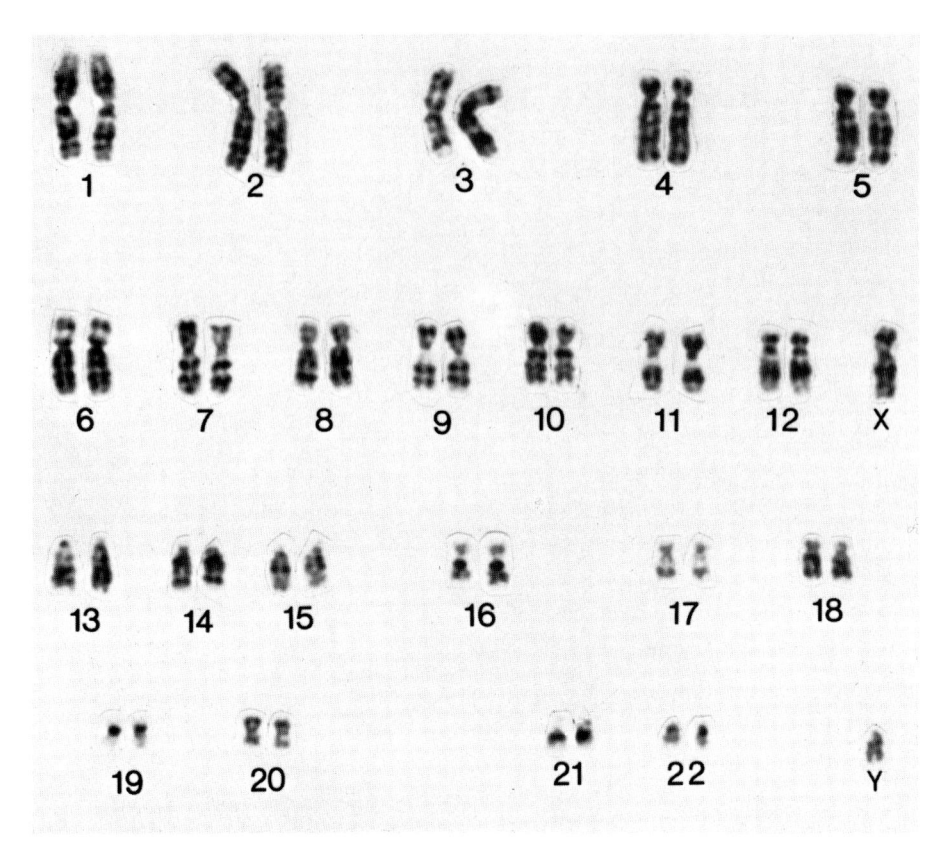

Fig. 5.5. G-band karyotype of man. [Sumner et al., Nature New Biology (London) *232*, 31–32 (1971)]

secondary constriction region in certain metacentric or submetacentric chromosomes, e.g., in human chromosomes 1, 9, and 16, and in the gorilla chromosomes 17 and 18, while other species do not. The first of these two findings suggests that man and the orangutan resemble each other more closely than either does the other species by lacking telomeric C-band regions, although it is well documented that man and the orangutan are in fact phylogenetically further apart than man is from the African apes (Chap. 10). The second tends to suggest that man and the gorilla may resemble each other more closely, not only by the fact that large C-banded secondary constrictions are present, but by the fact that they may occur in chromosomes recognized as homologues between the two species, such as HSA 16 and GGO 17. The significance of these findings is unclear, but as will be shown in Chap. 15, a distinction must be made between telomeric and secondary constriction C-bands in the Hominidae.

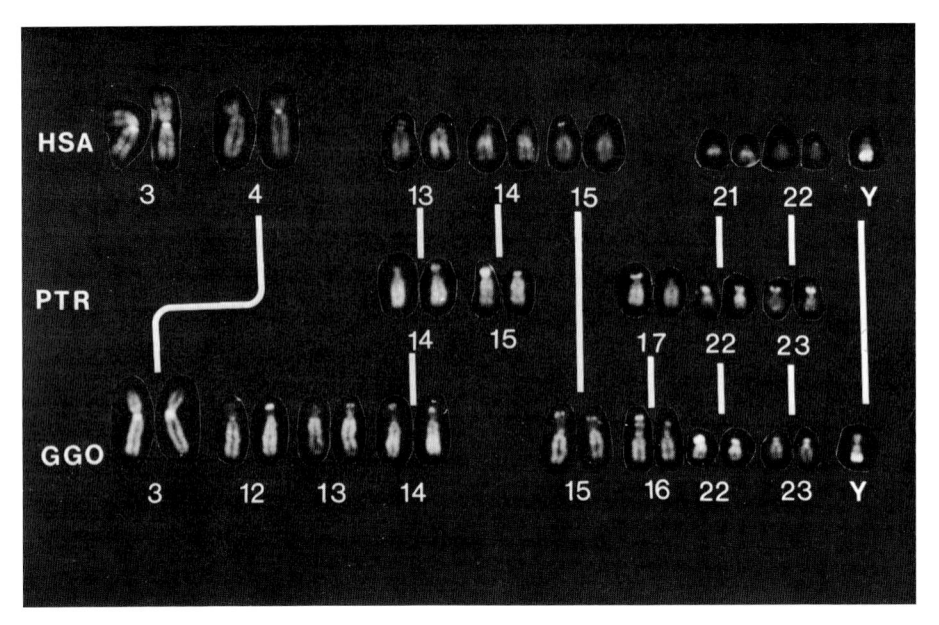

Fig. 5.6. Brilliant fluorescent regions in the chromosome complement of man (HSA), the chimpanzee, spp. *troglodytes*, (PTR), and the gorilla (GGO), with quinacrine staining. *Vertical bars* denote presumptive chromosome homologies between species. [Seuánez et al. Cytogenetics Cell Genetics *17*, 317–326 (1976)]

Telomeric C-bands correspond the positive Q-band regions that do not seem to contain homologous DNA sequences to any of the four major human satellite DNAs. C-band secondary constriction regions, on the other hand, are negatively Q-banded, and both in man and the gorilla they are sites of hybridisation with at least one cRNA to the four human DNA satellites. Finally, an interstitial C-band regions is found in the long arm of chromosome 6 and 14 in *Pan troglodytes* and chromosome 6 in *Pan paniscus*.

4. G-11 Staining

Bobrow et al. (1972) and Gagné and Laberge (1972) described a method for the demonstration of segments of human constitutive heterochromatin using Giemsa staining at a highly alkaline pH (G-11 technique). This procedure is especially useful in demonstrating the heterochromatic regions of chromosome 9 in man, although the technique also stains other heterochromatic regions of the human chromosome complement. When used to stain the chromosomes of *Pan troglodytes*

(Bobrow and Madan, 1973; Lejeune et al., 1973), *Gorilla gorilla* (Bobrow et al., 1972; Dutrillaux et al., 1973; Pearson, 1973), and *Pongo pygmaeus* (Dutrillaux et al., 1975c), these species showed larger regions of G-11 staining than man. Since it has been postulated that the technique demonstrates satellite III DNA-rich regions in man and in the great apes (Bobrow and Madan, 1973; Jones, 1976), the relationship between banding and the distribution of satellite III DNA in man and homologous sequences in the great apes will be analysed in detail in Chap. 15.

5. Methylated DNA Sequences

A different approach to the study of the nature of constitutive heterochromatin in man and the great apes has been reported by O.J. Miller et al. (1974) and Schnedl et al. (1975). This method is based on the property of 5-methyl-cytosine to become detectable by the technique of immunofluorescence. Antisera to 5-methylcytosine capable of reacting with 5-methylcystydylic residues in single-stranded DNA was used on chromosome preparations which had been denatured by UV light or by heat treatment. In man, the sites of antisera binding correspond to some regions of constitutive heterochromatin of the human chromosome complement in chromosome 1, 9, 15, 16, and the Y chromosome (O. J. Miller et al., 1974). Schnedl et al. (1975) compared two species of great apes *(Pan troglodytes* and *Gorilla gorilla)* with man, and it was evident that man and *Gorilla* appeared to be more closely related to one another than either was to the chimpanzee. The *Gorilla* showed large areas of antibody binding in chromosome 12, 13, 14, 15, 17, 18, and in the Y chromosome, but no comparable regions were detected in the chromosome complement of *Pan troglodytes,* where the general binding level was low.

6. T-Banding

These bands may be produced by heat denaturation, using a phosphate buffer solution followed by decoloration and subsequent staining with the fluorescent dye acridine orange (Dutrillaux, 1973). The majority of regions stained by this technique are located at the telomeres of the chromosome arms in man together with some intercalary bands in chromosome 11, 19, and 22, and they probably represent the most resistant

R-band regions to heat denaturation. In the two species of chimpanzee and the gorilla, the distribution of T-band regions is very similar to that in man, whereas in the orangutan, a larger number of regions were detected and their distribution was not strictly coincident (Dutrillaux, 1975; et al., 1975c). Although T-bands may be useful in demonstrating telomeric regions, especially in the case of translocations when studying human populations (see Dutrillaux, 1973), it is doubtful whether any evolutionary implication can be obtained by comparing T-band regions between man and the great apes. Dutrillaux (1975) has suggested that the short-arm region of PPY 10, which is strongly T-banded, may correspond to the region q11 and q22 in HSA 7, which consists of R-bands with some similar characteristics to T-bands. He has also proposed that the T-band region in PPY 8 which is located at the telomere of its short arm, might have been relocated to the subcentromeric region of HSA 11 by a pericentric inversion. However, a comparison between the G-band patterns of PPY 8 and HSA 11 strongly argues against the possibility that a pericentric inversion has taken place between these two chromosomes (Paris Conference, 1971; supplement 1975).

7. Ammoniacal Silver (Ag-AS) Staining

This technique is useful in demonstrating active sites of transcription of 18S and 28S rDNA cistrons, and will be discussed in detail in Chap. 16.

The Inference of Chromosome Homology Through Different Degrees of Similarity

One of the first questions which we may ask is how to evaluate the vast amount of information which can be obtained by different banding and staining techniques. Obviously, when different species are compared only the X chromosome of man and the great apes appears to have remained not only morphologically identical, but with identical Q-, G-, R-, C-, and T-banding. Other chromosomes may resemble each other with one technique, while differing from each other when studied by another technique. For example, chromosome 21 in man may resemble chromosome 22 in the two chimpanzees and the gorilla more closely than chromosome 22 in the orangutan, by showing a brilliant fluorescent region. However, with C-banding, chromosome 22 in the African apes

Benirschke, K., Bogart, M.H., McClure, H., Nelson-Rees, W.A.: Fluorescence of the trisomic chimpanzee chromosomes. J. Med. Primatol. *3*, 311–314 (1974)

Bobrow, M.: New techniques in chromosome cytology. In: Chromosome variations in human evolution. Boyce, A.J. (ed.), pp. 1–15. London: Taylor and Francis Ltd. 1975

Bobrow, M., Madan, K.: A comparison of chimpanzee and human chromosomes using the giemsa 11 and other chromosome banding techniques. Cytogenet. Cell Genet. *12*, 107–116 (1973)

Bobrow, M., Madan, K., Pearson, P.: Staining of some specific regions of human chromosomes, particularly the secondary constriction of No. 9. Nature New Biol. *238*, 122–124 (1972)

Bogart, M.H., Benirschke, K.: Chromosomal analysis of the pygmy chimpanzee *(Pan paniscus)* with a comparison to man. Folia Primatol. *27*, 60–67 (1977a)

Bogart, M.H., Benirschke, K.: Q-band polymorphism in a family of pygmy chimpanzees *(Pan paniscus)*. J. Med. Primatol. *6*, 172–175 (1977b)

Borgaonkar, D.S., Sadasivan, G., Ninan, T.A.: *Pan paniscus* Y chromosome does not fluoresce. J. Hered. *62*, 245–246 (1971)

Caspersson, T., Farber, S., Foley, G.E., Kudynowski, J., Modest, E.J., Simonsson, E., Wagh, V., Zech, L.: Chemical differentiation along metaphase chromosomes. Exp. Cell. Res. *49*, 219–222 (1968)

Chiarelli, B.: Chromosomes of the orangutan *(Pongo pygmaeus)*. Nature (London) *192*, 285 (1961)

Chiarelli, B.: Comparative morphometric analysis of the primate chromosomes. I. The chromosomes of the anthropoid apes and of man. Caryologia *15*, 99–121 (1962)

Chiarelli, B.: Le morfologia del cromosoma "Y" delle differenti specie di primati. Riv. Antropol. *54*, 137–140 (1967)

Chiarelli, B., Lin, C.C.: Comparison of fluorescent patterns in human and chimpanzee chromosomes. Genen Phaenen *15*, 103–106 (1972)

Chu, E.H.Y., Bender, M.A.: Cytogenetics and Evolution in primates. Ann. N. Y. Acad. Sci. *102*, 253–265 (1962)

Dutrillaux, B.: Nouveau systeme de marquage chromosomique: les bandes T. Chromosoma (Berlin) *41*, 395–402 (1973)

Dutrillaux, B.: Sur le nature el l'origine des chromosomes humaines. Paris: L'expansion Scientifique 1975

Dutrillaux, B., Lejeune, J.: Sur une nouvelle technique d'analyse du caryotype humain. C. R. Acad. Sci. (Paris) *272*, 2638–2640 (1971)

Dutrillaux, B., Rethoré, M.O., Aurias, A., Goustard, M.: Analyse du caryotype de deux espèces de gibbon *(Hylobates lar* et *Hylobates concolor)* par différents techniques de marquage. Cytogenet. Cell Genet. *15*, 81–91 (1975a)

Dutrillaux, B., Rethoré, M.O., Lejeune, J.: Analyse du caryotype de *Pan paniscus:* comparaison avec les autres Pongidae et l'homme. Humangenetik *28*, 113–119 (1975b)

Dutrillaux, B., Rethoré, M.O., Lejeune, J.: Comparaison du caryotype de l'orangutan *(Pongo pygmaeus)* a celui de l'homme du chimpanzee et du gorille. Ann. Genet. *18*, 153–161 (1975c)

Dutrillaux, B., Rethoré, M.O., Prieur, M., Lejeune, J.: Analyse de la structure fine des chromosomes du gorilla *(Gorilla gorilla)*. Comparaisons avec *Homo sapiens* et *Pan troglodytes*. Humangenetik *20*, 343–354 (1973)

Egozcue, J.: Primates. In: Comparative mammalian cytogenetics. Benirschke, K. (ed.), pp. 357–389. Berlin, Heidelberg, New York: Springer 1969

Egozcue, J., Aragonés, J., Caballín, R.M., Goday, C.: Banding patterns of the chromosomes of man and gorilla. Ann. Genet. *16*, 207–210 (1973a)

Egozcue, J., Caballín, R., Goday, C.: Banding patterns of the chromosomes of man and the chimpanzee. Humangenetik *18*, 77–80 (1973b)

Ford, C.E., Hamerton, J.L.: The chromosomes of man. Nature (London) *178*, 1020–1023 (1956)

Gagné, R. Laberge, C.: Specific cytological recognition of the heterochromatic segment of number 9 chromosome in man. Exp. Cell Res. *73*, 239–242 (1972)

Gallimore, P.H., Richardson, C.R.: An improved banding technique exemplified in the karyotype analysis of two strains of rat. Chromosoma (Berlin) *41*, 259–263 (1973)

Goodman, M.: Protein sequence and immunological specificity. Their role in phylogenetic studies of the primates. In: Phylogeny of the primates. Luckett, W.P., Szalay, J.S. (eds.), pp. 219–248. New York: Plenum Press 1975

Grouchy, de J., Turleau, C., Roubin, M., Chavan-Colin, F.: Chromosomal evolution of man and primates. In: Chromosome identification, techniques and applications in biology and medicine. Caspersson, T., Zech, L. (eds.), pp. 124–131. New York, London: Academic Press 1973

Grouchy, de J., Turleau, C., Roubin, M., Klein, M.: Evolutions caryotypiques de l'homme et du chimpanzé. Etude comparative des topographies des bandes après dénaturation ménagée. Ann. Genet. *15*, 79–84 (1972)

Hamerton, J.L., Fraccaro, M., de Carli, L., Nuzzo, F., Klinger, H.P., Hullinger, L., Taylor, A., Lang, E.M.: Somatic chromosomes of the gorilla. Nature (London) *192*, 225–228 (1961)

Hamerton, J.L., Klinger, H.P., Mutton, D.E., Lang, E.M.: The somatic chromosomes of the Hominoidea. Cytogenetics *2*, 240–263 (1963)

Jones, K.W.: Repetitive DNA sequences in animals. Proceedings of the Leiden Chromosome Conference (1974). In: Chromosomes today. Pearson, P.L., Lewis, K.R. (eds.), Vol. 5, pp. 305–313. New York: John Wiley & Sons. Jerusalem: Israel Univ. Press 1976

Khudr, G., Benirschke, K., Sedwick, C.J.: Man and *Pan paniscus:* A karyologic comparison. J. Hum. Evol. *2*, 323–331 (1973)

Klinger, H.P., Hamerton, J.L., Mutton, D., Lang, E.M.: The chromosomes of the Hominoidea. In: Classification and human evolution. Washburton, S.L. (ed.), pp. 235–242. Chicago: Aldine Publishing Co. 1963

Lejeune, J., Dutrillaux, B., Rethoré, M.O., Prieur, M.: Comparaison de la structure fine de chromatides d'*Homo sapiens* et de *Pan troglodytes*. Chromosoma (Berlin) *43*, 423–444 (1973)

Lin, C.C., Chiarelli, B., Boer, L.E.M. de, Cohen, M.M.: A comparison of fluorescent karyotypes of the chimpanzee *(Pan troglodytes)* and man. J. Hum. Evol. *2*, 311–321 (1973)

McClure, H.M., Belden, K.H., Pieper, W.A.: Autosomal trisomy in a chimpanzee: resemblance to Down's syndrome. Science *165*, 1010–1011 (1969)

McClure, H.M., Belden, K.H., Pieper, W.A., Jacobson, C.B., Picciano, D.: Cytogenetic studies and observations in the Yerkes great ape colony. In: Medical primatology; 1970 Proc. 2nd Conf. Exp. Med. Surg. Primates, (New York 1969), pp. 281–296. Basel: Karger 1971

Miller, D.A., Firschein, I.L., Dev, V.G., Tantravahi, R., Miller, O.J.: The gorilla karyotype, chromosome length and polymorphisms. Cytogenet. Cell Genet. *13*, 536–550 (1974)

Miller, O.J., Schnedl, W., Allen, J., Erlanger, B.F.: 5-methylcytosine localized in mammalian constitutive heterochromatin. Nature (London) *251*, 636–637 (1974)

Paris Conference (1971); Supplement 1975. Standardization in human cytogenetics Birth Defects: Original article series. Vol. XI, 9. New York: National Foundation 1975

Pearson, P.L.: The uniquenes of the human karyotype. In: Chromosome identification techniques and application in biology and medicine. Caspersson, T., Zech, L. (eds.), pp. 145–151. New York, London: Academic Press 1973

Pearson, P.L., Bobrow, M., Vosa, C.G., Barlow, P.N.: Quinacrine fluorescence in mammalian chromosomes. Nature (London) *231*, 326–329 (1971)

Schnedl, W., Dev, V.G., Tantravahi, R., Miller, D.A., Erlanger, B.F., Miller, O.J.: 5-methylcytosine in heterochromatic regions of chromosomes: chimpanzee and gorilla compared to the human. Chromosoma (Berlin) *52*, 59–66 (1975)

Seth, P.K., de Boer, L.E.M., Saxena, M.B., Seth, S.: Comparison of human and non-human primate chromosomes using fluorescent benzimidozal and other banding techniques. In: Chromosomes today. Pearson, P.L., Lewis, K.R. (eds.), Vol. 5, pp. 315–322. New York: John Wiley & Sons, Jerusalem: Israel Univ. Press 1976

Seuánez, H.: Chromosomes and spermatozoa of the great apes and man. Thesis, Univ. Edinburgh (1977)

Seuánez, H., Fletcher, J., Evans, H.J., Martin, D.E.: A chromosome rearrangement in an orangutan studied with Q-, C-, and G-banding techniques. Cytogenet. Cell Genet. *17*, 26–34 (1976a)

Seuánez, H., Fletcher, J., Evans, H.J., Martin, D.E.: A polymorphic structural rearrangement in the chromosomes of two populations of orangutan. Cytogenet. Cell Genet. *17*, 327–337 (1976b)

Stockholm Conference (1977) An International System for Human Cytogenetic Nomenclature, 1978. Birth Defects: Original Article Series. The National Foundation, in press

Tantravahi, R., Dev, V.G., Firschein, I.L., Miller, D.A., Miller, O.J.: Karyotype of the gibbons *Hylobates lar* and *Hylobates moloch*. Inversion in chromosome 7. Cytogenet. Cell Genet. *15*, 92–102 (1975)

Tjio, H., Levan, A.: The chromosome number of man. Hereditas *42*, 1–6 (1956)

Turleau, C., de Grouchy, J.: New observations on the human and chimpanzee karyotypes: identification of breakage points of pericentric inversions. Humangenetik *20*, 151–157 (1973)

Turleau, C., de Grouchy, J., Chavin-Colin, F., Mortelmans, J., Van den Bergh, W.: Inversion péricentrique du 3, homozygote et hétérozygote, et translation centremérique du 12 dans une famille d'orangs-outans: implication évolutives. Ann. Genet. *18*, 227–233 (1975)

Turleau, C., de Grouchy, J., Klein, M.: Phylogenie chromosomique de l'homme et des primates hominiens *(Pan troglodytes, Gorilla gorilla* and *Pongo pygmaeus)* essai de reconstitution du caryotype de l'ancêtre commun. Ann. Genet. *15*, 225–240 (1972)

Yeager, C.H., Painter, T.S., Yerkes, R.M.: The chromosomes of the chimpanzee. Science *91*, 74–75 (1940)

Young, W.J., Mergt, T., Ferguson-Smith, M.A., Johnston, A.W.: Chromosome number of the chimpanzee *(Pan troglodytes)*. Science *131*, 1672–1673 (1960)

Warburton, D., Firschein, I.L., Miller, D.A., Warburton, F.E.: Karyotype of the chimpanzee, *Pan troglodytes* based on measurements and banding patterns: comparison to the human karyotype. Cytogenet. Cell Genet. *12*, 453–461 (1973)

Chapter 6 Chromosome Heteromorphisms in Man and the Great Apes as a Source of Chromosome Variation Within Species

Chromosome Heteromorphisms in Man

Man is the best studied of all living beings as far as the number of cytologically examined individuals are concerned. Extensive population surveys have shown that chromosome aberrations, which could be divided into structural and numerical, occur with an incidence of 2.11% in man (Hamerton et al., 1975). The great majority of these chromosome aberrations correspond to clinical conditions that result from abnormal phenotypes in the carriers. From an evolutionary point of view, most chromosome aberrations represent situations where only individuals with reduced fitness are produced, whose survival would ultimately depend on medical treatment. Thus, it is highly unlikely that the vast majority of chromosome aberrations described in man could represent a source of chromosome variation within our species from which karyotypic evolution could eventually occur.

Some structural aberrations may occur, however, in individuals with a normal phenotype, such as some Robertsonian translocations between two acrocentric chromosomes of the human chromosome complement. Numerical aberrations are in very few cases compatible with a normal phenotype; XYY men fall into this rare category, although many of them have a remarkably abnormal social behaviour (Price et al., 1976). It must be pointed out, however, that even when carriers of structural or numerical chromosome aberrations appear phenotypically normal, the concept of "normality" applies only as far as no manifest clinical condition is detected in them. The possibility exists that metabolic disorders not yet clinically evident are present, or that the condition would predispose carriers to be more susceptible to certain diseases. Moreover, carriers of structural chromosome aberrations, though presumably normal, frequently produce chromosomally unbalanced offspring. Most reports in the literature of such carriers are ascertained through family studies where there has been a chromosomally abnormal proband.

In spite of the generally reduced fitness of these phenotypically normal carriers, some aberrant chromosomes have been found to be transmitted through many generations. Edwards et al. (1970), traced a D-G Robertsonian translocation for five generations, whereas Rott et al. (1972) traced a C-11/D-13 translocation for four, and Therkelsen (1971) a C-F translocation for five generations. Wahrman et al. (1972) reported that a pericentric inversion of chromosome 9 could be traced in five generations, and a small Y chromosome was traced in an Old Order Amish family for 11 generations (Borgaonkar et al., 1969), while another small Y chromosome was transmitted for at least 400 years in a French Canadian family (Genest and Lejeune, 1972). An inverted Y chromosome was found in an Indian family by Grace et al. (1972) for four generations, without apparently affecting the fertility of the male carriers.

When variant chromosomes are analysed in relation to their banding patterns an important distinction must be made. When change affects euchromatic regions the rule seems to be that these variant chromosomes, although they may occasionally be transmitted by phenotypically normal individuals for many generations, are of low incidence in human populations and frequently restricted to families. On the other hand, when change affects heterochromatic regions (such as those demonstrated by C-banding) or brilliant fluorescent regions, the variant chromosomes are generally carried by phenotypically normal individuals, and their incidence in the population is higher than that expected by recurrent mutation, thus conforming to the definition of a chromosome polymorphism. Here again, from an evolutionary point of view, it is only these latter that are capable of spreading within populations, and so becoming the source of new chromosome types from which karyotypic diversity may eventually occur. In man it has been estimated, for example, that approximately 1.5–18% of males show differences in the length of the Y chromosome (Soudek, 1973). Cohen et al. (1966) reported that Japanese men had a significantly higher Y chromosome length when compared to Indians, Negroes, Jews and non-Jews, while non-Jews showed the smallest Y chromosome size of the four groups tested. These results indicated that chromosome variants might occur with different incidence among ethnic groups, as was later reported by Lubs and Ruddle (1971). With the development of chromosome-banding techniques it was found that the size variation of the Y chromosome long arm in man was due to size variations of its brilliant fluorescent region (Bobrow et al., 1971), a region that could be either absent, or if present, be of different sizes (Fig. 6.1). In Australian aborigines, for example, the long arm of the Y chromosome is much reduced in size and the brilliant fluorescent region may even be absent (Angell, 1973). On the other hand, seven human autosomes (Nos. 3, 4, 13, 14, 15, 21, and 22) show variable regions when

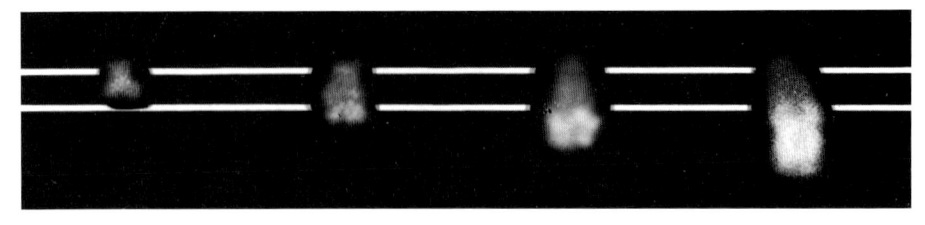

Fig. 6.1. Variable size of the brilliant long-arm distal region in the human Y chromosome. Courtesy of Professor H. John Evans

studied with Q-banding. Autosomal Q-band polymorphisms consist of variations in size and/or of fluorescent intensity between homologues, as well as among individuals. The satellites of all human acrocentric chromosomes, their proximal short-arm region, except for that of pair No. 14 and 15 and the near centromere region of chromosome 3 and 4, show brilliant polymorphisms in man (Buckton et al., 1976). Moreover, the subcentromeric brilliant region of chromosome 3 may occasionally be relocated in the proximal region of the short arm of this chromosome, probably as a result of an inversion. These kinds of fluorescent polymorphisms have been found to be stable within individuals, and chromosomes carrying these variant regions usually undergo a normal Mendelian segregation, although some deviations from expected Mendelian ratios have been reported (Buckton et al., 1976). To the above-mentioned polymorphisms we must also add those which result from size variations of the satellite stalks, or nucleolar organizer region of the acrocentric chromosomes, which are palely or negatively stained with quinacrine fluorescence. The terminal satellites of the acrocentric chromosomes may be, on the other hand, totally absent, so that a negatively stained satellite region might either result from the fact that satellites are not present or that, if present, they fail to show up with quinacrine fluorescence.

Polymorphisms identified with C-banding techniques involve differences in the size of their heterochromatic region of the centromere, and of the secondary constriction region of chromosome 1, 9, and 16, together with the already mentioned terminal region of the Y chromosome long arm which is also positively C-banded. Each of these regions showed a wide range of change when different individuals were compared, with a continuous spectrum of C-band size within populations. In two autosomes (1 and 9) the C-band subcentromeric secondary constrictions may not only vary in size, but may be relocated in the proximal region of the short arm as a result of inversions (Figs. 6.2 and 6.3). In some instances

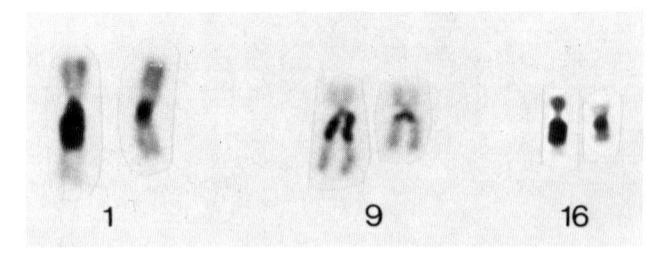

Fig. 6.2. Variation in size of the secondary constriction region of chromosome 1, 9, and 16 in man, with C-banding. Courtesy of Miss K. Buckton

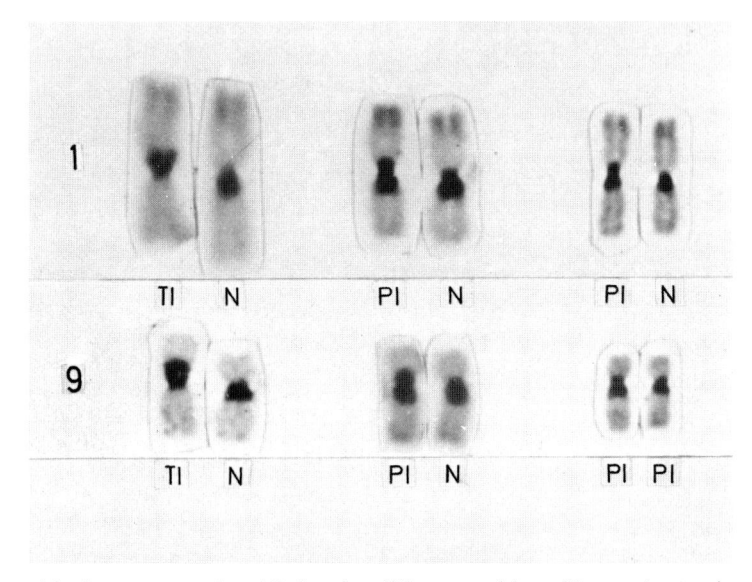

Fig. 6.3. Heteromorphic chromosomes 1 and 9 showing different position of heterochromatic region. *TI* total inversion; *PI* partial inversion; *N*, normal. [Buckton et al., Annals of Human Genetics (London) 40, 99–112 (1976)]

inversions may relocate the entire C-band region to the short arm, while in other cases only part of the heterochromatic region. These autosomal polymorphisms have also been found to segregate in a normal Mendelian fashion, although exceptions to this generalization have been reported as well (Robinson et al., 1976). The mechanism by which unequally sized C-band regions may be formed is still unclear; it has been postulated that they could result from unequal crossing over in meiotic or in somatic cells, giving rise to duplications and deficiencies at the C-band regions (Craig Holmes et al., 1975). C-band polymorphisms in the human Y chromosome have been extensively analysed with microdensitometric techniques (Drets and Seuánez, 1974). This study revealed that the heterochromatic long arm segment of the Y chromosome could be resolved into subregions characterized by densitometric peaks; their number being positively correlated with the relative size fo the whole C-terminal region of the Y chromosome (Fig. 6.4).

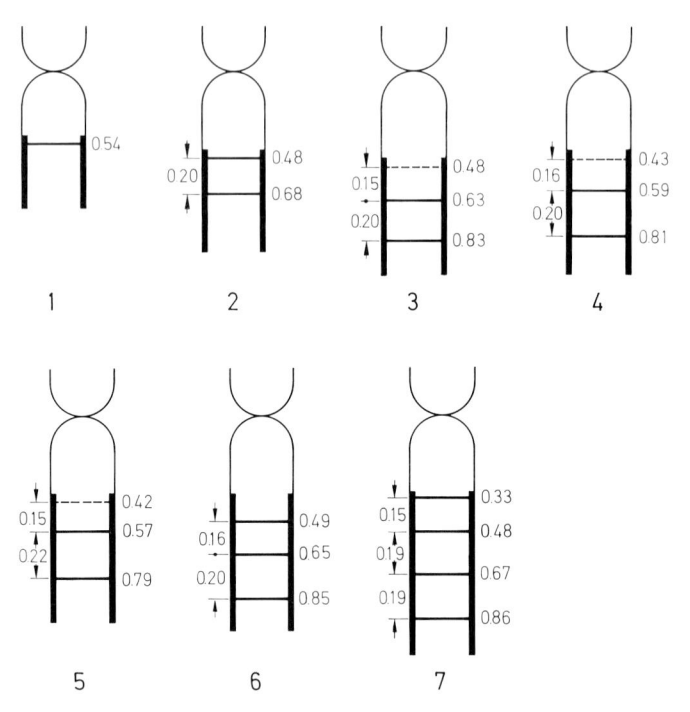

Fig. 6.4. Diagrammatic representation of the relative size of the heterochromatic segment and relative band position along the long arm of the Y chromosome. *1–7*, Individual Nos. Interband distances are indicated by *arrows* and *values*. *Dotted lines* denote those peaks detected only in a small number of chromosomes in persons classified as two-band carriers. Note the increase of interband distance and the similarity of interband values in different persons. [Drets and Seuánez, in Physiology and Genetics and Reproduction Vol. A, pp. 29–52; Ed. Coutinhno, E., and Fuchs, F., Plenum Press, New York (1974)]

However, the regions which are polymorphic in the human chromosome complement consist of apparently functionless chromatin, since they coincide with the sites where satellite DNA sequences have been located (Chap. 14). Since it is known that satellite DNAs are transcriptionally inactive, and moreover, since there is no proof that they may play a regulatory function in the genome, it is not clear what kind of selective advantage might result from the maintenance of these kinds of chromosome polymorphisms in human populations. On the other hand, if at any time selective advantage resulted from any of the observed chromosome polymorphisms it should be stressed that human populations are today no longer genetically isolated from each other. We are a highly mobile species, and extensive migration and interbreeding make it very unlikely that any kind of new karyotype could ever become fixed in a population and subsequently lead to speciation.

Chromosome Heteromorphisms in the Great Apes

Contrary to the abundance of material made available for cytogenetic surveys in human pupulations, the number of specimens of great ape studied with chromosome banding techniques is remarkably low. Not only are great ape species endangered in their natural environment, but the few specimens in zoos and research centres are not always available for experimentation, a reason why relatively few animals have been studied. Thus, most reports published are based on chromosome analysis of very few individuals, the highest number of specimens of chimpanzee reported in a single publication accounting for 22 (Lin et al., 1973), of pygmy chimpanzees for 8 (Bogart and Benirschke, 1977), of gorillas for 16 (Seuánez, 1977), and of orangutans for 27 (Seuánez, 1977).[2] Nonetheless, much has been gained from the reports, as well as from other studies based on very small numbers of undividuals.

1. Chromosome Heteromorphisms in Pan troglodytes

In this species brilliant Q-band polymorphisms have been found at the proximal region of the short arm of the acrocentric chromosomes Nos. 14, 15, 17, 22, and 23, but not at their terminal satellites as might happen in some of their human homologue chromosomes (Lin et al., 1973). As in human autosomal Q-band polymorphisms they account for variations in size and/or in the intensity fo quinacrine staining; size variations of brilliant fluorescent regions in eight specimens are shown in Fig. 6.5. The frequency of brilliant polymorphisms in the chimpanzee is considerably higher than in man; in man their incidence averages from about 2.9–4.2 (Buckton et al., 1976), while in the chimpanzee it has been estimated to be equal to 8.77 (Lin et al., 1973), and to 8.85 (Seuánez, 1977). The palely or negatively stained satellite stalks of these acrocentric chromo-

---

[2] A report on 23 out of these 27 animals had already been published (Seuánez et al., 1976b).

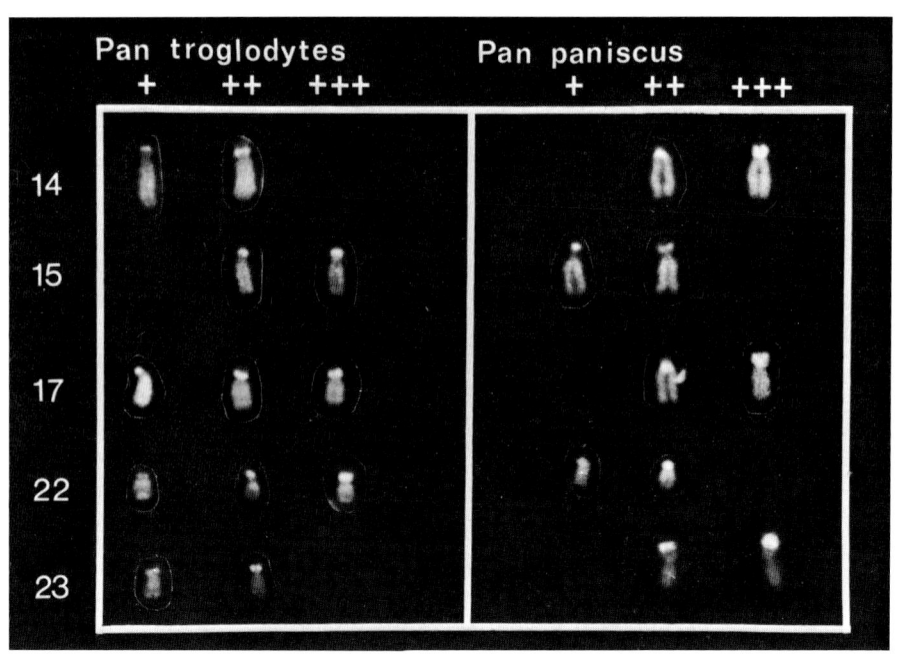

Fig. 6.5. Size variations of brilliant fluorescent regions in *Pan troglodytes* and *Pan paniscus*

somes are also variable in size; their variability can, however, be better demonstrated with techniques for visualizing the nucleolar organizer regions of these chromosomes (Chap. 16). Q-band polymorphisms are also evident at the telomeres of many chromosomes of the chimpanzee complement. These regions are stained with medium intensity, and variation may then be only in size, but not in their staining intensity. These regions are better demonstrated with C-banding.

C-band polymorphisms in *Pan troglodytes* account for variations in the size of constitutive heterochromatin regions at the centromere of all chromosomes, the very small short-arm region of Y chromosome, as well as size variations of the telomeric Q-, C-band regions. Interstitial C-band regions (PTR 6 and 14) are probably variable as well.

2. Chromosome Heteromorphisms in Pan paniscus

This species is very similar to *Pan troglodytes,* and brilliant Q-band polymorphisms have been found at the proximal region of the short arm of the acrocentric chromosomes 14, 15, 17, and 21, as well as at the short arm of the small metacentric pair No. 23. Variations account for changes in size and/or intensity of quinacrine staining. Figure 6.5 shows size variations of brilliant fluorescent regions observed in three specimens (Seuánez, 1977). Bogart and Benirschke (1977) reported a family of pygmy chimpanzees in which a female adult animal had a deletion of the short arm of chromosome 23, thus lacking the brilliant fluorescent region. This variant chromosome had been transmitted to six of her offspring. As in *Pan troglodytes*, *Pan paniscus* shows telomeric Q-, C-band regions at many chromosome arms; with Q-banding these regions are stained with medium intensity and are variable only in size.

C-band polymorphisms are similar to those observed in *Pan troglodytes*, except that chromosome 23 in the pygmy chimpanzee shows a large heterochromatic block which comprises the entire arm of this small metacentric chromosome, a region which is also stained intensely with G-11 techniques (Dutrillaux et al., 1975a). The interstitial C-band region in chromosome 6 is probably variable as well.

3. Chromosome Heteromorphisms in Gorilla gorilla

In this species, autosomal brilliant Q-band polymorphisms have been found in the centromere region of chromosome 3, the terminal satellites of the acrocentric chromosomes 12, 13, 14, 15, and 16, and at the proximal short arm regions of chromosome 22 and 23 (see Figs. 6.6 and 6.7). Variations account for change in size and/or fluorescent intensity (D. A. Miller et al., 1974). The frequency of brilliant autosomal polymorphisms accounted for 14.9 in ten specimens of gorilla, which is approximately five times the number observed in man (Seuánez, 1977). In two animals of this group we have found a short-arm deletion in chromosome 14, so that the satellite region was absent. Brilliant fluorescence is found at the Y chromosome long-arm distal tip in this species; on the other hand, we failed to observe any size variation of this region in eight male specimens. Q-band polymorphisms are also evident at the telomeric region of many chromosome arms (Fig. 6.6); these regions which are stained with medium intensity may vary in size between homologue chromosomes and among individuals and can be better demonstrated with C-banding.

C-band polymorphisms are evident at (1) the centromere region of all chromosome pairs except GGO 11, where there is no C-band at the centromere, (2) at the secondary constriction region of chromosome 17 and 18, (3) at positively C-banded satellite stalk regions of chromosomes 12, 13, 14, 15, and 16 that may vary in length, and (4) at the terminal (telomeric) C-band regions (Fig. 6.6).

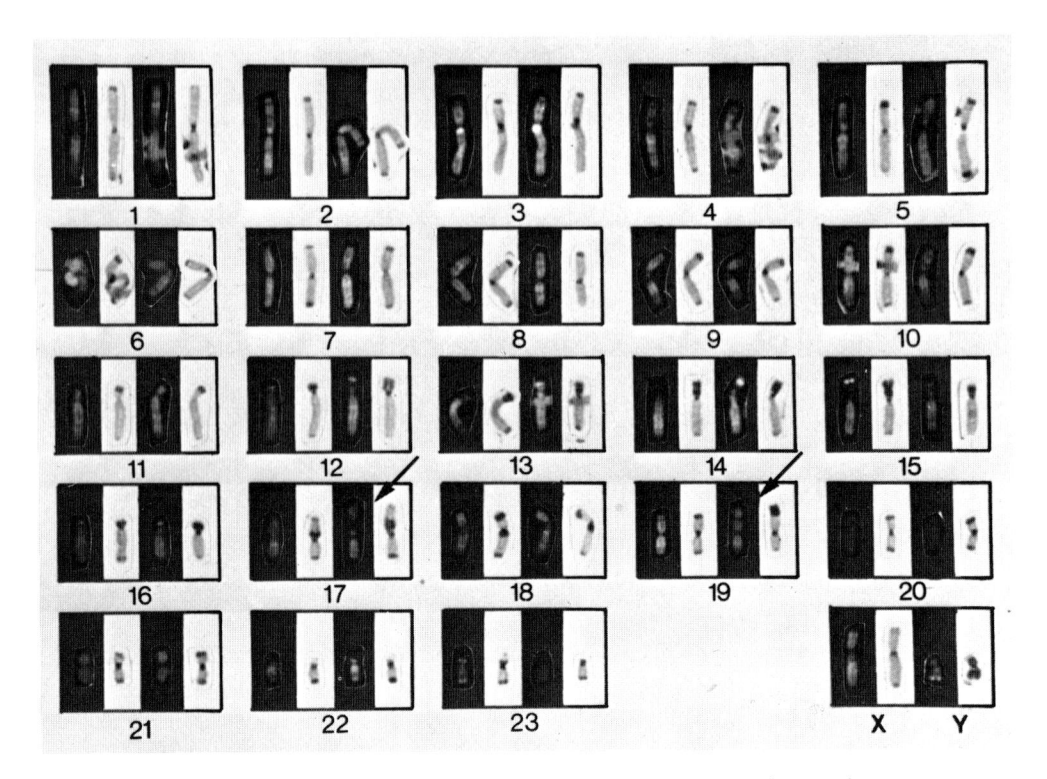

Fig. 6.6. Q-, C-chromosome polymorphisms in the gorilla. *Arrows* point to chromosomes showing large heterochromatic regions

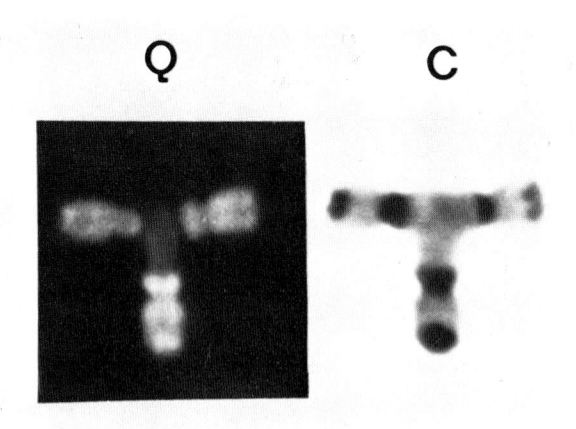

Fig. 6.7. Q–C-band patterns of chromosomes 22–23 in the gorilla. Note that only the proximal region of the short arm may be intensely or brilliantly stained with quinacrine, but the terminal satellites are negatively stained

4. Chromosome Heteromorphisms in Pongo pygmaeus

In this species brilliant fluorescence is not found in any region of the chromosome complement and brilliant Q-band polymorphisms are thus non-existent. Some variable regions may, however, show up with quniacrine, e.g., the short-arm region of chromosome 23 seen in one specimen, which was of greater size than that of its normal homologue in the same animal. This large short arm was, however, better demonstrated by C-banding (Seuánez et al., 1976a); it consisted of a large heterochromatic block which transformed this chromosome into a metacentric chromosome. C-band polymorphisms account for differences in the size of the heterochromatic region of the centromere of all chromosomes except for PPY 9 which shows no C-band region at its centromere (Fig. 15.3). The Y chromosome shows a large heterochromatic region below the centromere and a smaller one in the long-arm distal tip (Fig. 15.5).

Chromosome analysis of 29 specimens of orangutan have revealed interesting findings in intraspecific chromosome variation. In the first place, the animals studied could be divided into three different groups in terms of origin: Bornean (B), Sumatran (S), and hybrids between the two populations (H). In the pedigrees of Fig. 6.8, all the adult animals (B.I and S.I) had been caught wild, while all the young animals (B.II, S.II, and H.1) had been bred in captivity. All adult animals except B.I.3 and B.I.4, which had been recently mated, and S.I.15, which had never mated, were of proven fertility (Seuánez, 1977). Chromosome analysis showed variant chromosomes No. 2, 9, 14, and 22. Two different kinds of chromosome 2 were observed in these animals; one carried by all the Bornean animals studied (9 specimens) in the homozygous condition consisted of a subtelocentric chromosome 2 with a centromeric index=10.2, with one G-band region in its short-arm and two G-band regions in the subcentromeric region (Fig. 6.9). From pedigree studies this chromosome could be traced in eight more Bornean animals not studied by us (Fig. 6.8). In all the animals of Sumatran origin studied (16 specimens), the homologue to this chromosome corresponded to a more metacentric chromosome 2 (centromeric index=18.9), with two G-band regions in its short-arm and one in the subcentromeric region (Fig. 6.9). The two types of chromosome 2 were also compared with Q- and R-banding, and the patterns observed confirmed the results obtained with G-banding. Such a comparison suggested that either type of chromosome 2 could be derived from the other by a pericentric inversion. Finally, the four hybrid animals studied were carriers of both types of

Fig. 6.8. Orangutan pedigree

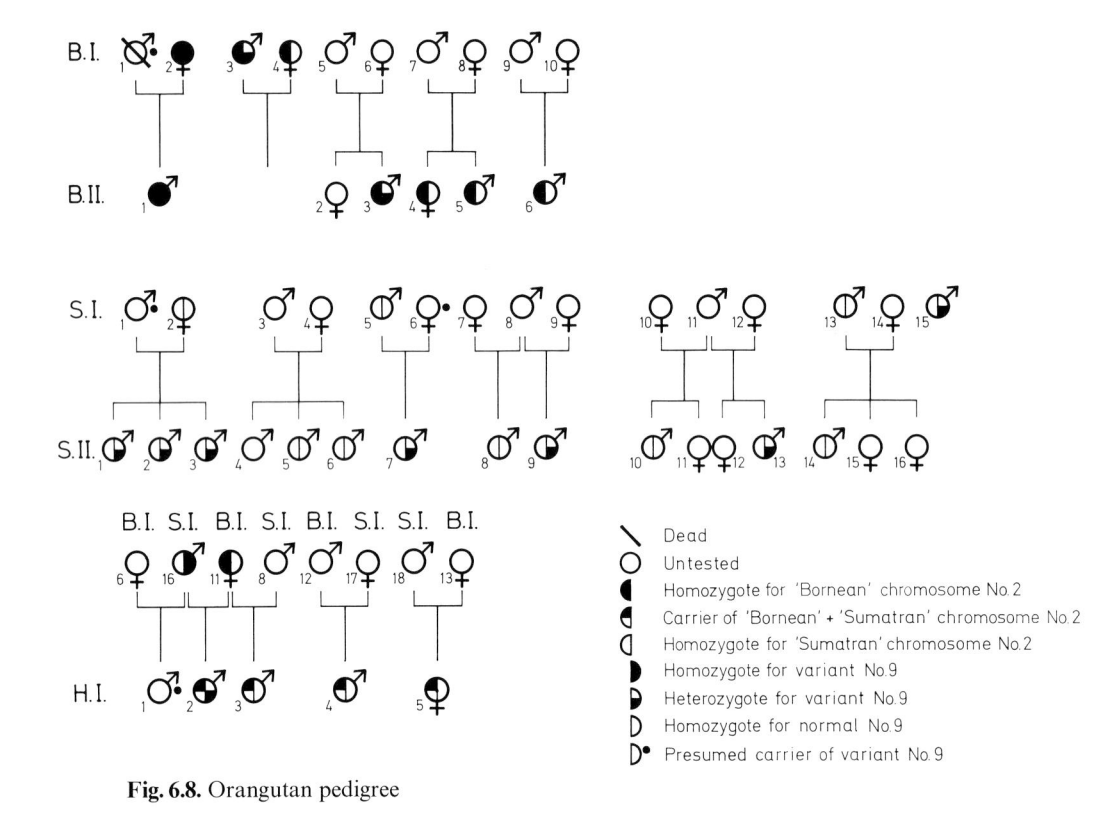

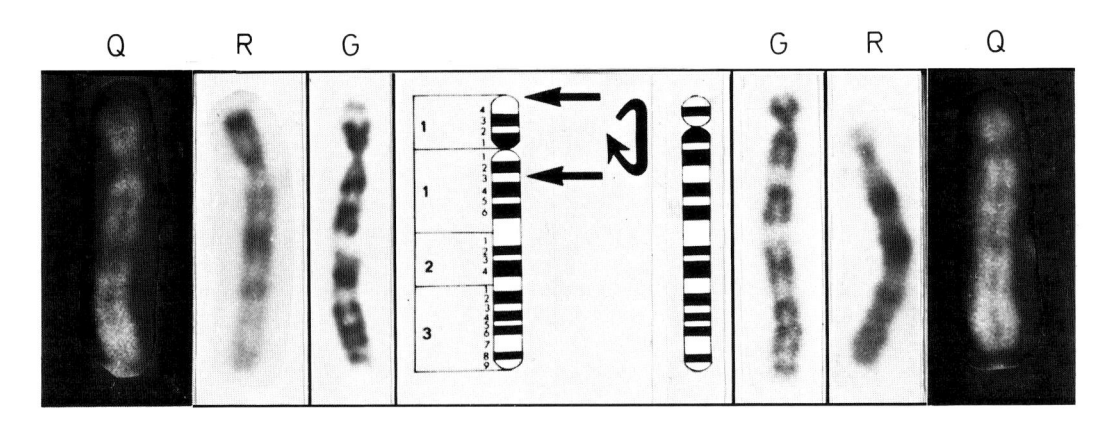

Dead
Untested
Homozygote for 'Bornean' chromosome No. 2
Carrier of 'Bornean' + 'Sumatran' chromosome No. 2
Homozygote for 'Sumatran' chromosome No. 2
Homozygote for variant No. 9
Heterozygote for variant No. 9
Homozygote for normal No. 9
Presumed carrier of variant No. 9

Fig. 6.9. Q-, R-, and C-banding patterns of the Sumatran *(left)* and Bornean *(right)* chromosome 2 in the orangutan. The Bornean chromosome can be derived from the Sumatran by a pericentric inversion *(curved arrow)* between two chromosome breaks *(straight arrows)*. [Seuánez et al., Cytogenetics Cell Genetics 23, 137–140 (1979)]

chromosome 2. In two of them (H.I.2 and H.I.3) it was possible to know from which parent (Bornean or Sumatran) each type of chromosome 2 had been transmitted. One of these hybrid animals (H.I.2) was, in addition, a heterozygote carrier of a variant chromosome 9 which had been transmitted by his father (S.I.16).

A variant chromosome 9 was found in 13 animals, 10 heterozygous and 3 homozygous carriers, and was traced in four other carriers not studied by us (Fig. 6.8). The variant chromosome was present in both Bornean and Sumatran orangutans, and the three possible combinations (normal PPY 9 carriers, heterozygous carriers and homozygous carriers of the variant chromosome) were found in each subpopulation. The variant chromosome could not be simply derived from the normal chromosome 9; it was a more metacentric chromosome with a centromeric index=34.7 against 19.2 of its normal homologue. With G-banding (Fig. 6.10) it was demonstrated that the variant chromosome could be derived from the normal by a complex rearrangement interpreted as an inversion within an inversion (Seuánez et al., 1976b), a finding also confirmed by Q- and R-banding. Q- and C-banding studies showed no C-band region at the centromere of the variant chromosome 9.

The variant chromosomes 14 and 22 consisted of chromosomes in which the whole short-arm region had been deleted. One animal (H.I.3) carried these two deletions and was also heterozygote for the two types of chromosome 2. A deleted chromosome 22 was also found in his mother (B.I.11), and in a sib (H.I.2), who was also a carrier of both types of chromosome 2 and 9.

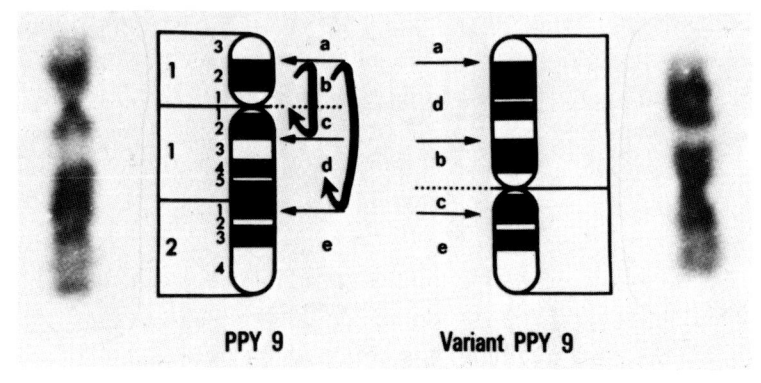

Fig. 6.10. G-banding pattern of the normal *(left)* and the variant *(right)* chromosome 9 in the orangutan. The variant chromosome *(inverted)* can be derived from the normal by two inversions *(curved arrows)*, one inside the other. Break points are indicated by *straight arrows*. The rearranged chromosome can be described as pter→p13::q21→q13::p13→q13::q21→qter. [Seuánez et al., Cytogenetics Cell Genetics 17, 327–337 (1976)]

Phylogenetic Implications of Chromosome Variation in the Orangutan

Of the four kinds of variant chromosomes analysed, two of them, 14 and 22, occurred in one and in three related individuals, respectively, whereas the other two, 2 and 9, were found in many specimens. The variant chromosomes produced by deletions accounted for loss of the short arm, satellite stalks, and terminal satellite regions of these two chromosomes; they therefore implied loss of 18S and 28S rDNA sequences (Chap. 16), as well as of highly repetitive DNA sequences homologous to human satellite I, II, and III (Chap. 14). The other two cases deserve to be analysed in detail, since they are different from one another and occur independently of each other in this species. In the first place, the distribution observed in chromosome 2 is good evidence that the incidence of each chromosome type differs significantly between orangutan subpopulations, suggesting that each chromosome type has become fixed in each subpopulation of orangutan following selection, genetic drift, and perhaps a high degree of inbreeding as well. For this reason we have proposed (Seuánez, 1977; Seuánez et al., 1979) that each chromosome type should be designated as the "Bornean" and "Sumatran" chromosome 2 in *Pongo pygmaeus* rather than referring to them as "normal" and "variant" chromosomes. Since, as we have previously stated, each type of chromosome 2 can be derived from the other by one pericentric inversion, the question arises which of these can be considered the ancestral form. Some light on this question comes from a comparison with other species of great ape and man. Chromosome 2 in *Pongo pygmaeus* is homologous to chromosome 2 in *Pan troglodytes* (PTR 2), *Pan paniscus* (PPA 2), and *Gorilla gorilla* (GGO 2), and to chromosome 3 in man (Chap. 5). On the principle of parsimony, which assumes that a network of descent can be best explained with the fewest number of changes, the ancestral chromosome 2 in the orangutan is more likely to be that from which we can most simply derive PTR 2, PPA 2, GGO 2, and HSA 3. It is possible to derive these chromosomes from the "Bornean" chromosome 2 with only two breaks and one pericentric inversion (Fig. 6.11), whereas more than two breaks would be needed to derive them from the "Sumatran" chromosome 2. This suggests that the "Bornean" chromosome 2 might be the original chromosome of the species, from which the "Sumatran" chromosome type has been derived by pericentric inversion.

These observations are of interest on two grounds. First they show that the pericentric inversion involving euchromatic regions is a mecha-

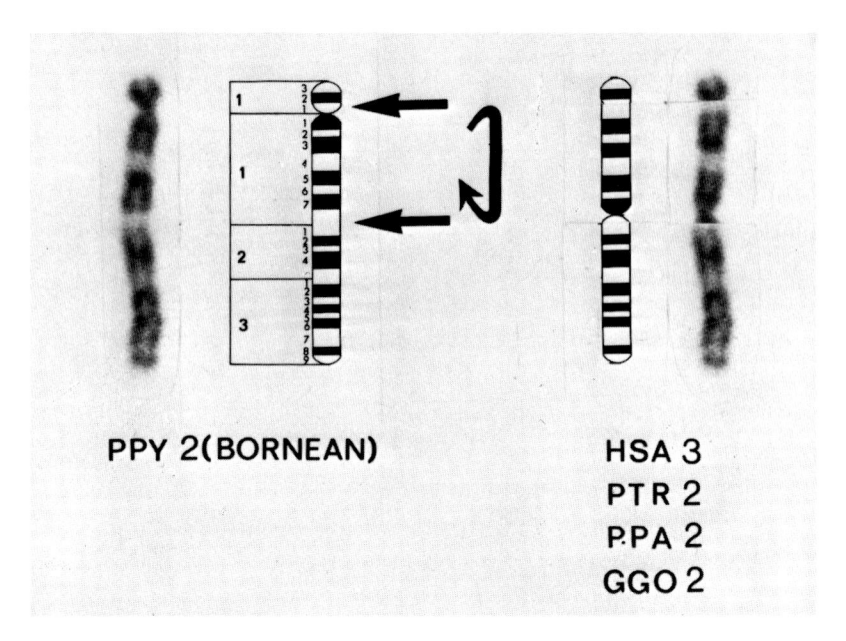

Fig. 6.11. Derivation of chromosome 3 in man (HSA 3) and 2 in the chimpanzee (PTR 2), the pygmy chimpanzee (PPA 2) and the gorilla (GGO 2) from the Bornean chromosome 2 in the orangutan by a single pericentric inversion. [Seuánez et al., Cytogenetics Cell Genetics 23, 137–140 (1979)]

nism by which new chromosome types have been formed in *Pongo pygmaeus,* and support the view that such kind of inversion was of importance in the evolution of man and the great apes from a common stock (see Bobrow and Madan, 1973; Turleau et al., 1972; de Grouchy et al., 1973). Second, they have implications in relation to the preservation of the orangutan, a species which is at present endangered in its natural environment, with population sizes of approximately 3000 in Borneo and 1000 in Sumatra (Napier and Napier, 1967). The possibility of producing hybrids by mating captive animals from the two islands in zoos and research centres must be most carefully considered, since many such matings would produce offspring heterozygous for one pericentric inversion. There is already one report published in which this event has been demonstrated (Turleau et al., 1975) in addition to the examples noted in our present studies. The hybrid animals studied by us, which were bred in captivity, were prepubertal, and it will be of primary importance to assess their fertility in the future. Since the inversion is pericentric and involves about one fifth of the total chromosome length, the probability of chiasma formation within an inversion pairing loop – which could result in unbalanced gametes – may be quite low. However, we have suggested (Seuánez et al., 1979) that it might be prudent to avoid matings between Bornean and Sumatran orangutans in captivity, and to maintain separate breeding programmes for the two types.

Variant chromosomes 9 in *Pongo pygmaeus* have been described (although they have often been designated with different criteria of nomenclature) by Lucas et al. (1973), Dutrillaux et al. (1975b), Turleau et al. (1975), Seuánez et al. (1976a and b), and Seuánez (1977). A detailed study of these reported variant chromosomes indicates that they all represent an identical rearrangement involving chromosome 9. This must therefore be a common type of variant chromosome with a high incidence in the population of *Pongo pygmaeus,* obviously higher than what could be maintained by recurrent mutation, thus conforming to the definition of a chromosome polymorphism. Since the variant chromosome is present in the two subpopulations of orangutan, the complex chromosome rearrangement giving rise to this variant chromosome 9 must have occurred before the original population of orangutan spread into Sumatra and Borneo, a period which goes back to at least 8000 years. Both subpopulations became, since then, completely isolated from each other by the South China Sea. Thus, the fact that this rearrangement is of ancient origin and has a widespread distribution suggests that it has been maintained in both subpopulations as a balanced polymorphism. For this to happen, we would except that some advantage must exist for heterozygous carriers, although this is difficult to prove. We have commented (Seuánez et al., 1976b) that this rearrangement might, in the heterozygous state, result in the formation of two inversion loops at the first meiotic division (Fig. 6.12), and only if crossing-over were confined to the terminal segments of the paired chromosomes would balanced gametes be produced. We know that in the C-group chromosomes in man, among which we find the homologous chromosome to PPY 9 (HSA 12), most chiasmata are found at the terminal region of the chromosome arms (Hulten, 1974). If this were the case in *Pongo,* and if chiasma position is an accurate reflection of cross-over site, then this preferential location of chiasmata might well contribute to a maintained fertility of the heterozygous carriers. There is a least one report in the literature (Turleau et al., 1975) in which an heterozygous animal for a variant chromosome 9 was mated to a normal homozygous carrier for chromosome 9, and the variant chromosome was transmittet to two offspring.

In mammals, generally, Robertsonian translocations are frequently found as chromosome polymorphisms (White, 1973), but other types of polymorphic chromosome rearrangements are less frequent. There

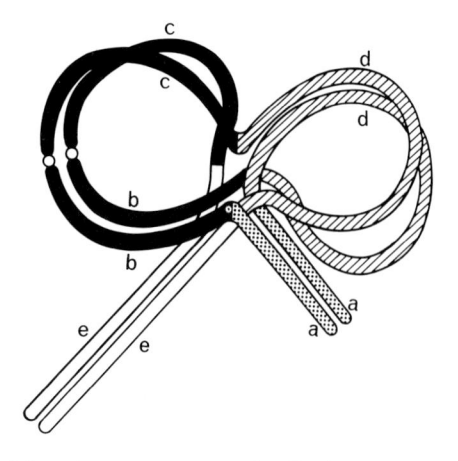

Fig. 6.12. Theoretical configuration of bivalent 9 in a heterozygous carrier. Regions *a* to *e* are the regions illustrated in Figure 6.10. Cross-overs at regions *a* and *e* will result in balanced products. A single cross-over at region *b* or *c* would result in unbalanced products of the duplication-deficiency type. A single cross-over at region *d* would result in unbalanced products, carrying dicentric chromosomes and acentric fragments. [Seuánez et al., Cytogenetics Cell Genetics *17*, 327–337 (1976)]

is evidence, however, that pericentric inversions may be polymorphic in populations of the deer mouse (Ohno et al., 1966; Arakaki et al., 1970); plains wood rat (Baker et al., 1970); South American populations of *Rattus rattus* (Bianchi and Paulette-Vanrel, 1969); mountain voles (Gileva and Pokrovski, 1970); and African pygmy mice (Matthey and Jotterand, 1970, Jotterand, 1972).

In man, pericentric inversions involving whole or partial C-band regions occur with relatively high frequencies in at least two chromosomes in the complement (Buckton et al., 1976). However, these human inversions are small, and appear to be confined to regions occupied by constitutive heterochromatin—such regions being considered of low genetic activity—and are not therefore comparable to the situation described here in *Pongo pygmaeus*. To be sure, there are reports of larger pericentric inversions in man which involve euchromatic chromosome regions, but these are often associated with pathological conditions. In some less common cases, these larger pericentric inversions have been found to be transmitted by phenotypically normal heterozygotes who showed normal fertility (Jacobs et al., 1967; Weitkamp et al., 1969; Crandal and Sparkes, 1970; Betz et al., 1974; de la Chapelle et al., 1974; Jacobs et al., 1974), but the incidence of chromosomes with such inversions in human populations is extremely low. Moreover, all cases were ascertained through some kind of clinically abnormal condition, and in only one

(Betz et al., 1974) was the inversion present in the homozygous state. Thus in man, the best-studied of all hominoids, none of the pericentric inversions that involve significant amounts of euchromatic material can be considered to be a fixed balanced polymorphism in the population.

References

Angell, R.R.: The chromosomes of Australian Aborigines. In: The human biology of Aborigines in Cape York. Kirk, R.L. (ed.), pp. 103–109. Canberra: A.C.T. 1973

Arakaki, D.T., Veomett, I., Sparkes, R.S.: Chromosome polymorphism in deer mouse siblings *(Peromyscus maniculatus)*. Experientia 24, 425–426 (1970)

Baker, J.R., Mascarello, J.T., Jordan, R.G.: Polymorphism in the somatic chromosomes of *Neotoma micropus Baird*, the plains woodrat. Experientia 26, 426–428 (1970)

Betz, A., Turleau, C., Grouchy, de J.: Hétérozygotie et homozygotie pour une inversion péricentrique du 3 humain. Ann. Genet. 17, 77–80 (1974)

Bianchi, N.O., Paulette-Vanrel, J.: Complement with 38 chromosomes in two South American populations of *Rattus rattus*. Experientia 25, 1111–1112 (1969)

Bobrow, M., Madan, K.: A comparison of chimpanzee and human chromosomes using the Giemsa 11 and other chromosome banding techniques. Cytogenet. Cell Genet. 12, 107–116 (1973)

Bobrow, M., Pearson, P.L., Pike, M.S., El-Alfi, O.S.: Length variation in the quinacrine binding segment of the Y chromosomes of different sizes. Cytogenetics 10, 190–198 (1971)

Bogart, M.H., Benirschke, K.: Q-band polymorphism in a family of pygmy chimpanzees *(Pan paniscus)*. J. Med. Primatol. 6, 172–175 (1977)

Borgaonkar, D.S., McKusick, U.A., Herr, H.M., de los Cobos, L., Yoder, O.C.: Constancy of the length of the Y chromosome. Ann. Genet. 12, 262–264 (1969)

Buckton, K.E., O'Riordan, M.L., Jacobs, P.A., Robinson, J.A., Hill, R., Evans, H.J.: C- and Q-band heteromorphisms in the chromosomes of three human populations. Ann. Hum. Genet. (London) 40, 99–112 (1976)

Chapelle, de la A., Schröder, J., Stenstrand, K., Fellman, J., Herva, R., Saarni, M., Anttolainen, I., Tallila, I., Tervilä, L., Husa, L., Tallqvist, G., Robson, E.B., Cook, P.J.L., Sanger, R.: Pericentric inversions of human chromosomes 9 and 10. Am. J. Hum. Genet. 26, 746–766 (1974)

Cohen, M.M., Shaw, M., MacCluer, J.W.: Racial differences in the length of the human Y chromosomes. Cytogenetics 5, 34–52 (1966)

Craig-Holmes, A.P., Moore, F.B., Shaw, M.W.: Polymorphism of human C-band heterochromatin II. Family studies with suggestive evidence for somatic crossing-over. Am. J. Hum. Genet. 27, 178–179 (1975)

Crandall, B.F., Sparkes, R.S.: Pericentric inversion of a number 15 chromosome in nine members of a family. Cytogenetics 9, 307–316 (1970)

Drets, M.E., Seuánez, H.: Quantitation of heterogeneous human heterochromatin: microdensitometric analysis of C- and G-bands. In: Physiology and genetics of reproduction. Coutinho, E., Fuchs, F. (eds.), Vol. I, pp. 29–52. New York: Academic Press 1974

Dutrillaux, B., Rethoré, M.O., Lejeune, J.: Analyse du caryotype de *Pan paniscus:* comparaison avec les autres Pongidae et l'homme. Humangenetik 28, 113–119 (1975a)

Dutrillaux, B., Rethoré, M.O., Lejeune, J.: Comparaison du caryotype de l'orangutan *(Pongo pygmaeus)* a celui de l'homme du chimpanzé et du gorille. Ann. Genet. 18, 153–161 (1975b)

Edwards, J.A., Finley, A.J., Marinello, M.J.: A large kindred with D-G translocation mongolism. J. Med. 11, 193–200 (1970)

Genest, P, Lejeune, J.: Recherche sur l'origine d'un petit chromosome Y multicentenaire. Ann. Genet. 15, 51–53 (1972)

Gileva, E.A., Pokrovsky, A.V.: Karyotype characteristics and chromosome polymorphism in Pamir Alay mountain voles of the *Microtus juldaschi* (Cricetidae) group. Zool. Zh. *49*, 1229–1239 (1970)

Grace, H.J., Ally, F.E., Paruk, M.A.: 46, X inv. (Yp$^+$, q$^-$) in four generations of an Indian family. J. Med. Genet. *9*, 293–297 (1972)

Grouchy, de J., Turleau, C., Roubin, M., Chavan-Colin, F.: Chromosomal evolution of man and primates. In: Chromosome identification, techniques and applications in biology and medicine. Caspersson, T., Zech, L. (eds.), pp. 124–131. New York, London: Academic Press 1973

Hamerton, J.L., Canning, C., Ray, M., Smith, S.: A cytogenetics survey of 14,069 newborn infants. Clin. Genet. *8*, 223–243 (1975)

Hulten, M.: Chiasma distribution at diakinesis in the normal human male. Hereditas *76*, 55–78 (1974)

Jacobs, P.A., Buckton, K.E., Cunningham, C., Newton, M.: An analysis of the break points of structural rearrangements in man. J. Med. Genet. *11*, 50–64 (1974)

Jacobs, P.A., Chruickshank, G., Faed, M.J.W., Robson, E.B., Harris, H., Sutherland, I.: Pericentric inversion of a group C autosome: a study of three families. Ann. Hum. Genet. (London) *31*, 219–230 (1967)

Jotterand, M.: Le polymorphisme chromosomique des *Mus* (Leggadas) *africains*. Cytogénétique, zoogeographie, évolution. Rev. Suisse Zool. *79*, 287–359 (1972)

Lin, C.C., Chiarelli, B., Boer, de L.E.M., Cohen, M.M.: A comparison of fluorescent karyotypes of the chimpanzee *(Pan troglodytes)* and man. J. Hum. Evol. *2*, 311–321 (1973)

Lubs, H.A., Ruddle, F.H.: Chromosome polymorphism in American negro and white populations. Nature (London) *233*, 134–136 (1971)

Lucas, M., Page, C., Tanmer, M.: Chromosomes of the orangutan, Jersey Wildlife Preservation Trust. Ann. Rep. 1973, pp. 57–58 (1973)

Matthey, R., Jotterand, M.: A new system of polymorphism non-Robertsonian in "Leggadas" *(Mus sp)* of the Centrafrican Republic. Rev. Suisse Zool. *77*, 630–635 (1970)

Miller, D.A., Firschein, I.L., Dev, V.G., Tantravahi, R., Miller, O.J.: The gorilla karyotype, chromosome length and polymorphisms. Cytogenet. Cell Genet. *13*, 536–550 (1974)

Napier, J.R., Napier, P.H.: A handbook of living primates. London, New York: Academic Press 1967

Ohno, S., Weiler, C., Poole, J., Christian, L., Stenius, C.: Autosomal polymorphism due to pericentric inversions in the deer mouse *(Peromyscus maniculatus)* and some evidence of somatic segregation. Chromosoma (Berlin) *18*, 177–187 (1966)

Price, W.H., Brunton, M., Buckton, K., Jacobs, P.A.: Chromosome survey of new patients admitted to the four maximum security hospitals in the United Kingdom. Clin. Genet. *9*, 389–398 (1976)

Robinson, J.A., Buckton, K.E., Spowart, G., Newton, M., Jacobs, P.A., Evans, H.J., Hill, R.: The segregation of human chromosome polymorphisms. Ann. Hum. Genet. (London) *40*, 113–122 (1976)

Rott, H.D., Schwanitz, G., Gross, K.P., Alexandrov, G.: C11/D13 translocation in four generations. Humangenetik *14*, 300–305 (1972)

Seuánez, H.: Chromosomes and spermatozoa of the great apes and man. Thesis, Univ. Edinburgh (1977)

Seuánez, H., Fletcher, J., Evans, H.J., Martin, D.E.: A chromosome rearrangement in an orangutan studied with Q-, C-, and G-banding techniques. Cytogenet. Cell Genet. *17*, 26–34 (1976a)

Seuánez, H., Fletcher, J., Evans, H.J., Martin, D.E.: A polymorphic structural rearrangement in the chromosomes of two populations of orangutan. Cytogenet. Cell Genet. *17*, 327–337 (1976b)

Seuánez, H., Evans, H.J., Martin, D.E., Fletcher, J.: An inversion of chromosome 2 that distinguishes between Bornean and Sumatran orangutans. Cytogenet. Cell Genet. *23*, 137–140 (1979)

Soudek, D.: Chromosomal variants with normal phenotype in man. J. Hum. Evol. *2*, 341–355 (1973)

Therkelsen, A.J.: A family with a presumptive CF translocation in 5 generations. Ann. Genet. *14,* 13–21 (1971)

Turleau, C., Grouchy, de J., Chavin-Colin, F., Mortelmans, J., Van den Bergh, W.: Inversion péricentrique du 3, homozygote et hétérozygote, et translation centromérique du 12 dans une famille d'orangs-outans: implications évolutives. Ann. Genet. *18,* 227–233 (1975)

Turleau, C., Grouchy, de J., Klein, M.: Phylogenie chromosomique de l'homme et des primates hominiens *(Pan troglodytes, Gorilla gorilla* et *Pongo pygmaeus)* essai de reconstitution du caryotype de l'ancêtre commun. Ann. Genet. *15,* 225–240 (1972)

Wahrman, J., Atidia, J., Goitein, R., Cohen, T.: Pericentric inversions of chromosome 9 in two families. Cytogenetics *11,* 132–144 (1972)

White, M.J.D.: Animal cytology and evolution, 3rd edit. Cambridge: Cambridge University Press 1973

Weitkamp, L.R., Janzen, M.K., Guttormsen, S.A., Gershowitz, H.: Inherited pericentric inversion of chromosome number two: a linkage study. Ann. Hum. Genet. (London) *33,* 53–59 (1969)

Pan paniscus – Pongo pygmaeus: 4 pericentric inversions + one paracentric inversion

Gorilla gorilla – Pongo pygmaeus: 6 pericentric inversions + one paracentric inversion

The highest number of chromosome rearrangements (7) are those between *Gorilla* and *Pongo,* and the second highest is that between *Gorilla* and man (6). Very surprisingly, the latter have a higher number than man and *Pongo* (5), although there is clear evidence that *Pongo* has split from the Hominidae before the splitting of the African apes and man as estimated by immunological distances (Goodman, 1975) and experiments of in vitro DNA hybridisation (Hoyer et al., 1972; Beneviste and Todaro, 1976). The fact that man and *Pongo* show less chromosome change, and have retained plesiomorph characteristics, such as the absence of telomeric C-band regions (patristic relationship), apparently obscures the fact that man and *Gorilla* have diverged later from a common ancestor (cladistic relationship). This is obviously due to the fact that the rate with which chromosomal change has occurred has not been the same in all species; but the chromosomes of *Gorilla* diverged much more rapidly from the chromosomes of man than did the chromosomes of *Pongo,* although this species had separated earlier in time. A comparison between *Pan troglodytes* and *Pongo* shows a smaller number of chromosome rearrangements (5) than between *Gorilla* and *Pongo* (7), whereas only two rearrangements have occurred between *Pan troglodytes* and *Gorilla.* There is good evidence that these two species are more closely related to one another than any of them is to *Pongo* (Goodman, 1975; Hoyer et al., 1972; Beneviste and Todaro, 1976). Thus, the difference in the number of chromosome rearrangements suggests that the rate of chromosome change between *Pan troglodytes* and *Pongo* has been slower than between *Gorilla* and *Pongo.* The most logical explanation is that the chromosomes of *Pan troglodytes* have evolved less rapidly than those of *Gorilla,* thus retaining more plesiomorphic characteristics (patristic relationship). It is not surprising, then, that only four chromosome rearrangements have occurred between man and *Pan troglodytes,* whereas six have occurred between man and *Gorilla.* This patristic relationship between man and *Pan troglodytes* might in fact obscure the fact that man and *Gorilla* might be the most closely related species, which will be shown by sharing apomorph and not plesiomorph characteristics. Thus, it is important to establish which will be the apomorph characteristics of any pair of species, should they have emerged from a more recent common ancestor.

Man – *Pan troglodytes*:	None
Man – *Gorilla gorilla*:	1) Y chromosome with brilliant fluorescence
	2) One homologous pair with brilliant fluorescence at the centromere region (HSA 4 = GGO 3)
	3) Secondary constriction in one homologous chromosome pair (HSA 16 = GGO 17)
	4) 5-methylcytosine rich regions in 3 homologous chromosomes (HSA 15 = GGO 15; HSA 16 = GGO 17; HSA Y = GGO Y)[3]
Pan troglodytes – Gorilla gorilla:	1) Terminal Q-, C-bands
	2) Two pericentric inversions; one: HSA 5, PPY 4→PTR 4, PPA 4 and GGO 4; the other: HSA 12, PPY 9→PTR 10, PPA 10 and GGO 10.

Thus, it appears that either man and *Gorilla* or *Pan troglodytes* and *Gorilla* have diverged from a recent common ancestor in the Hominidae, but no recent common ancestor seems to have existed between man and *Pan troglodytes*. One indication that man and *Gorilla* might have a common ancestor comes from the fact that only these two species show brilliant fluorescence in the Y chromosome and one autosome pair, and moreover, at the same regions of these homologous chromosomes (the centromeric region of HSA 4 and GGO 3, and the distal long arm of the Y chromosome). A second indication is the appearance of secondary constrictions that have taken place exclusively in these two species, and in particular in one homologue chromosome, HSA 16, and GGO 17. A third is the fact that in these two species there are detectable amounts of 5-methylcytosine-rich DNA sequences which are distributed at the same chromosome sites. In man, methylated DNA sequences have been detected by immunofluorescence techniques at the C-band regions of chromosome 1, 9, 15, 16, and the Y chromosome (O.J. Miller et al., 1974). In *Gorilla gorilla* these sequences have been detected at the C-band region of chromosomes 12, 13, 14, 15, 17, 18, and the Y chromosome (Schnedl et al.,

---

[3] GGO 13 and HSA 9 have been found to be homologues with comparative gene mapping (Chap. 12). Since both GGO 13 and HSA 9 have methylcytosine rich regions the distribution of these sequences is at actually 4 homologous chromosomes.

1975), whereas in *Pan troglodytes* none of the chromosomes showed large detectable amounts of 5-methylcytosine-rich regions. Similar studies in *Pongo pygmaeus* are needed to clarify this point completely. This is why the absence of detectable amounts of these sequences in *Pan troglodytes* cannot be taken as proof in favour of a common ancestor man-*Gorilla gorilla*. However, what suggests than man and *Gorilla gorilla* might have had a common ancestor is rather the distribution of the 5-methylcytosine-rich regions in their karyotypes. The fact that these regions are located at many homologous chromosomes of man and *Gorilla gorilla* (HSA 15 and GGO 15; HSA 16 and GGO 17; HSA Y and GGO Y as well as between HSA 9 and GGO 13) indicates that the amplification of these sequences could have occurred in a common ancestor of both species. Otherwise we would have to postulate that the similar distribution of these DNA sequences in these two species has resulted from a random event.

Thus, the chromosomes of man and *Gorilla gorilla* may differ in the number of chromosome rearrangements that have occurred between them, but in spite of these changes the overall phyletic affinity between the karyotype of *Gorilla gorilla* and man may be greater than between *Pan troglodytes* and man (D. A. Miller, 1977; Seuánez, 1977) contrary to the generally held view that the chimpanzee is man's closest living relative.

References

Beneviste, R. E, Todaro, G. J.: Evolution of type C viral genes: evidence for an Asian origin of man. Nature (London) *261*, 101–108 (1976)

Dallapicolla, B., Mastroiacovo, P., Gandini, E.: Centric fission of chromosome No. 4 in the mother of two patients with trisomy 4p. Hum. Genet. *31*, 121–125 (1976)

Dutrillaux, B.: Sur le nature el l'origine des chromosomes humaines. Paris: L'expansion Scientifique 1975

Egozcue, J.: Primates. In: Comparative mammalian cytogenetics. Benirschke, K. (ed.), pp. 357–389. Berlin, Heidelberg, New York: Springer 1969

Egozcue, J.: A possible case of centric fission in a primate. Experientia *27*, 969–970 (1971)

Finaz, C., van Cong, N., Frézal, J., de Grouchy, J.: Histoire naturelle du chromosome 1 chez les primates. Ann. Genet. *20*, 85–92 (1977 a)

Finaz, C., van Cong, N., Frézal, C., de Grouchy, J.: Fifty million year evolution of chromosome 1 in the primates: evidence from banding and gene mapping. Cytogenet. Cell Genet. *18*, 160–164 (1977 b)

Fredga, K., Bergström, U.: Chromosome polymorphism in the root vole *(Microtus oeconomus)*. Hereditas *66*, 145–152 (1970)

Goodman, M.: Protein sequence and immunological specificity. Their role in phylogenetic studies of the primates. In: Phylogeny of the primates. Luckett, W. P., Szalay, J. S. (eds.), pp. 219–248. New York: Plenum Press 1975

Grouchy de, J., Finaz, C., van Cong, N.: Comparative banding and gene mapping in the primates. Evolution of chromosome 1 during fifty million years. In: Chromosomes today. de la Chapelle, A., Sorsa, M. (eds.), Vol. 6, pp. 183–190. Amsterdam: Elsevier/North Holland Biomedical Press 1977

Hansen, S.: A case of centric fission in man. Humangenetik 26, 257–259 (1975)

Hoyer, B. H., van de Velde, N. W., Goodman, M., Roberts, R. B.: Examination of Hominid Evolution by DNA Sequence Homology. J. Hum. Evol. 1, 645–649 (1972)

Hsu, L. Y. F., Kim, H. J., Sujansky, E., Kousseff, B., Hirshhorn, K.: Reciprocal translocation versus centric fusion between two No. 13 chromosomes. Cytogenet. Cell Genet. 12, 235–244 (1973)

Lau, Y. F., Arrighi, F. E.: Studies of the squirrel monkey, *Saimiri sciureus*, genome. I. Cytological characterizations of chromosomes heterozygosity. Cytogenet. Cell Genet. 17, 51–60 (1976)

Lejeune, J., Dutrillaux, B., Lafourcade, J., Berger, R., Abonyi, D., Rethoré, M. O.: Endoreduplication sélective du bras long du chromosome 2 chez une femme et sa fille. C. R. Acad. Sci. (Paris) 266, 24–26 (1968)

Lejeune, J., Dutrillaux, B., Rethoré, M. O., Prieur, M.: Comparison de la structure fine des chromatides d'*Homo sapiens* et de *Pan troglodytes*. Chromosoma (Berlin) 43, 423–444 (1973)

Ma, N. S. F., Jones, T. C., Thorington, R. W., Cooper, R. W.: Chromosome banding patterns in squirrel monkeys, *Saimiri sciureus*. J. Med. Primatol. 3, 120–137 (1974)

Martin, R. D.: The bearing of reproductive behaviour and ontogeny on strepsirhine phylogeny. In: Phylogeny of the primates. Luckett, W. P., Szalay, J. S. (eds.), pp. 265–297. New York and London: Plenum Press 1975

Miller, D. A.: Evolution of Primate Chromosomes. Science 198, 1116–1124 (1977)

Miller, O. J., Schnedl, W., Allen, J., Erlanger, B. F.: 5-methylcytosine localized in mammalian constitutive heterochromatin. Nature (London) 251, 636–637 (1974)

Niebuhr, E., Skovby, F.: Cytogenetic studies in seven individuals with an i(X9) karyotype. Hereditas 86, 121–128 (1977)

Paris Conference (1971); Supplement 1975. Standardization in human cytogenetics birth defects: original article series, XI, 9. New York: National Foundation 1975

Roubin, M., Grouchy de, J., Klein, M.: Les Felides: Evolution chromosomique. Ann. Genet. 16, 233–245 (1973)

Schnedl, W., Dev, V. G., Tantravahi, R., Miller, D. A., Erlanger, B. F., Miller, O. J.: 5-methylcytosine in heterochromatic regions of chromosomes: chimpanzee and gorilla compared to the human. Chromosoma (Berlin) 52, 59–66 (1975)

Seuánez, H.: Chromosomes and spermatozoa of the great apes and man. Thesis, Univ. Edinburgh (1977)

Simpson, G. G.: Recent advances in methods of phylogenetic inference. In: Phylogeny of the primates. Luckett, W. P., Szalay, J. S. (eds.), pp. 3–19. New York, London: Plenum Press 1975

Sinha, A. K., Pathak, S., Nova, J. J.: A human family suggesting evidence for centric fission and stability of a telocentric chromosome. Hum. Hered. 22, 423–429 (1972)

Sokal, R. R., Sneath, P. N. A.: Principles of numerical taxonomy. San Francisco: H. Freeman & Co. 1963

Tantravahi, R., Dev, V. G., Firschein, I. L., Miller, D. A., Miller, O. J.: Karyotpye of the gibbons *Hylobates lar* and *Hylobates moloch*. Inversion in chromosome 7. Cytogenet. Cell Genet. 15, 92–102 (1975)

White, M. J. D.: Animal cytology and evolution, 3rd Edition. Cambridge: Cambridge Univ. Press 1973

Yoshida, T. H.: Evolution of karyotypes and differentiation in 13 *Rattus* species. Chromosoma (Berlin) 40, 285–297 (1973)

Chapter 8 Chromosome Variation Versus Chromosome Fixation

Allopatric and Stasipatric Models of Speciation

As we have discussed in previous chapters, the phylogeny of human and great ape chromosomes can be explained by chromosome rearrangement; nevertheless we do not yet know whether such a change led to speciation or whether it took place after speciation. In this respect, a great controversy exists among evolutionary biologists on the roles of chromosome change and geographical isolation in speciation. On one side, Mayr (1963) has defined a species as a population separated from others by discontinuity, the three main isolating mechanisms producing such discontinuity being (1) geographical, (2) ecological, and (3) reproductive. Obviously the ultimate mechanism that keeps species as separate taxa is reproductive isolation (Chap. 4). Now, according to Mayr (1963) geographical isolation is the essential mechanism by which speciation is produced; the other isolating mechanisms being only secondarily developed. These secondary mechanisms by which species are kept as separate populations may operate at a later stage in the event that these species might be again brought into contact. They could act at the pre, intra, or post mating levels, but their main effect is to avoid cross-hybridisation between different species, so that each of them may conserve its own genetic pool. Thus, the role of chromosome change in speciation has been categorically denied by Mayr (1963), since chromosome change has occurred, in his opinion, after speciation was produced by geographical isolation. This model of speciation has been named allopatric, and it has been objected to on the grounds that some animal populations defined as species which have been geographically isolated for long periods may produce normal and fertile offspring if brought again into contact. This is the case of the polar bear *(Thalarctos maritimus)* and the brown bear *(Ursus arctos)*, or the red deer *(Cervus elaphus)* and the wapiti or elk *(Cervus canadiensis)* (Short, 1976). On the other side, White (1968, 1973) has suggested that speciation may follow a "stasipatric" model; i.e., that species may have emerged through chromosome change while sharing a common habitat. Initially, chromosome rearrangement, being an uncommon event, would necessarily result in the appearance of an heterozygous carrier for a rearranged chromosome within a population. If fertile, this carrier could transmit the variant chromosome to its offspring. If the rearrangement involved large euchromatic regions and consisted of a pericentric inversion, for example, two effects could result: one, that the order of genes in the variant chromosome would differ from that of the normal chromosome; two, that crossing over within the inversion loop at meiosis would result in unbalanced gametes. Any mutation occurring at the regions involved in the chromosome rearrangement would then be carried by the chromosome in which it occurred, and would not be recovered in a viable product if it passed to its homologous partner by crossing over. Thus, mutations would be accummulated by each chromosome type, and this would make them both morphologically and genetically different. If the incidence of heterozygous animals in the population were high enough to allow two of them to mate, homozygous carriers for the rearranged chromosome would appear in the offspring, and these would differ from the homozygous carriers of the "normal chromosome" both genetically and in chromosome constitution. Consequently, selection might favour of eliminate one of these two chromosome types by acting upon individuals that have accummulated different mutant substitutions and also differ in the order of genes for the chromosome pair involved in the structural rearrangement. As pointed out be Bodmer (1975), a change in the order of genes might result in an arrangement capable of providing selective advantage, so that a new order could then become fixed in the population. An important objection to this model is that infertility barriers might restrict chromosomal change to those individuals in which it initially occurred, especially in complex organisms such as mammals. Our findings in *Pongo pygmaeus,* in which one rearranged chromosome has been reported to be widespread in two subpopulations for at least 8000 years, are a clear indication that chromosomal change may not necessarily be restricted by infertility barriers in the heterozygous condition. However these findings do not prove that stasipatric speciation has occurred. Wilson et al. (1974a, b; 1975) have observed that in mammals the rate of chromosomal evolution is higher than in other taxa, and have proposed that this might be explained by the way mammalian species and populations are socially organized, so that a high amount of inbreeding and genetic drift may occur. In fact, both models of speciation, allopatric and stasipatric, may have contributed to the evolution of mammals. Arnason (1972) has compared the evolution of the Cetacea and Pinnipedia with that of the Insectivora and Rodentia, and concluded that the former have probably envolved following an allopatric model of speciation, while the latter have evolved following a stasipatric model. As a general rule for mammalian species, Arnason (1972) held that allopatric speciation must have occurred in those showing high karyotypic stability, which correspond to those having:

1) Low reproduction rate { Late sexual maturity
   Reduced litter size and few litters per year
2) Good mobility
3) Environment without delimited niches

On the other hand, stasipatric speciation applied well to species showing:

1) High reproduction rate { Early sexual maturity
   Large litters and many litters per year
2) Restricted mobility
3) Closely delimited ecological niches

Man and the great apes seem to correspond to the first group of mammals, and this in turn suggests that speciation has probably been allopatric rather than stasipatric. Our findings in two subpopulations of *Pongo pygmaeus* may be interpreted in favour of geographical isolation as necessary to precede fixed chromosome change. The fact that the two subpopulations differed in the frequency of one chromosome type (the Bornean and Sumatran chromosome 2) suggests that geographical isolation is needed so that genetic drift or selection may operate. On the other hand, the fact that a variant chromosome 9 is widespread in both subpopulations supplies good evidence that chromosome rearrangements do not lead to speciation per se. These observations, however, do not prove that allopatric speciation is either taking place in present populations of *Pongo pygmaeus,* or that it has occurred between man and the great apes.

Before concluding this chapter it must be remembered, however, that man seems to have gradually emerged as a distinct species while sharing a common habitat with other phylogenetically related hominid[4] species (Chap. 2). R.E.F. Leaky and Walker (1976) reported that two hominid forms of remarkably different degree of hominization (*Homo erectus* and *Australopithecus boisei*) coexisted side by side in the Lake Turkana region of East Africa. Comparable findings in Olduvai had indicated that *Homo habilis* and *Australopithecus boisei* had also been contempor-

---

[4] "hominid" is used to refer to fossil forms belonging to the human lineage after the phyletic divergence of man and the great apes from the common stock.

aneous species (L.S.B. Leaky et al., 1964), while von Koenigswald (1973) had also found evidence in Asia in favour of the existence of contemporaneous hominid lineages (Chap. 2). These findings critically question the "single-species hypothesis" by which the emergence of *Homo sapiens* had previously been explained. R. E. F. Leaky and Walker (1976) observe that this hypothesis was based on the assumption that culture and toolmaking were the primary adaptations of our direct hominid ancestors by means of which open country environments were conquered. It is also asserted that other basic attributes such as bipedal posture, reduced size of canines, and delayed physical maturity were developed in response to a greater cultural dependence. These assumptions coupled with the concept that the most advanced hominids had succeeded in displacing other less developed hominids by competitive exclusion led anthropologists to consider it unlikely that man could have emerged sympatrically.

If man then has evolved as a distinct species while sharing a common habitat with other less advanced hominids, it is obvious that the establishment of reproductive isolation between hominid populations must have depended on factors other than simple geographical isolation. Among these factors chromosome rearrangement could have played a significant role. It must be stressed, however, that although chromosome change could have occurred comparatively late in the lineage of man, namely after the split from the common stock with the great apes, the kinds of rearrangement we have discussed in previous chapters must have occurred at an earlier stage, if they were to give rise to different lineages of phyletic divergence leading to modern man and to modern apes.

References

Arnason, U.: The role of chromosomal rearrangement in mammalian speciation with special reference to Cetacea and Pinnipedia. Hereditas 70, 113–118 (1972)

Bodmer, W.F.: Analysis of linkage by somatic hybridisation and its conservation by evolution. In: Chromosome variations in human evolution. Boyce, A.T. (ed.), pp. 53–61. London: Taylor and Francis Ltd. 1975

Koenigswald, von G.H.R.: *Australopithecus, Maganthropus* and *Ramapithecus*. J. Hum. Evol. 2, 487–491 (1973)

Leaky, S.L.B., Tobias, P.V., Napier, J.R.: A new species of the genus *Homo* from Olduvai Gorge. Nature (London) 202, 7–9 (1964)

Leaky, R.E.F., Walker, A.C.: *Australopithecus, Homo erectus* and the single species hypothesis. Nature (London) 261, 572–574 (1976)

Mayr, E.: Animal species and evolution. Cambridge, Mass.: Harvard University Press 1963

Short, R.V.: The origin of species. In: Reproduction in mammals. The evolution of reproduction. Austin, C.R., Short, R.V. (eds.), Vol. 6, pp. 110–148. Cambridge: Cambridge Univ. Press 1976

White, M.J.D.: Models of speciation. Science *159*, 1065–1070 (1968)
White, M.J.D.: Animal cytology and evolution. 3rd ed. Cambridge: Cambridge Univ. Press 1973
Wilson, A.C., Maxson, L.R., Sarich, V.M.: Two types of molecular evolution. Evidence from studies of interspecific hybridisation. Proc. Nat. Acad. Sci. USA *71*, 2843–2847 (1974a)
Wilson, A.C., Sarich, V.M., Maxson, L.R.: The importance of gene rearrangement in evolution. Evidence from studies on rates of chromosomal protein and anatomical evolution. Proc. Nat. Acad. Sci. USA *71*, 3028–3030 (1974b)
Wilson, A.C., Bush, G.L., Case, S.M., King, M.C.: Social structuring of mammalian populations and rate of chromosomal evolution. Proc. Nat. Acad. Sci. USA *72*, 5061–5065 (1975)

Section III
Comparative Gene Mapping and Molecular Cytogenetics
A New Approach to Cytotaxonomy

Chapter 9 Composition of the Human Genome

Repetitive and Non-Repetitive DNA Sequences

One of the most intriguing findings that substantially changed our conception of the nature and function of the eukaryotic genome was the fact that a very substantial amount of DNA is composed of repetitive sequences, most of them apparently meaningless. When eukaryotic DNA was sheared into short-length segments, denatured and later reannealed under carefully controlled conditions, it was observed that the rate of renaturation followed a second-order reaction, or was a function of the product of the initial DNA concentration (in mol of nucleotides per litre) and the time of incubation (in s), or Co.t (Britten and Kohne, 1968). The observation that eukaryotic DNA was reassociating at different values of Co.t was explained as resulting from the degree of repetition within the genome. Reassociation of DNA needs the collision of two complementary single-stranded chains; the higher their concentration in the solution, the higher their probability of reannealing at a given time. If a DNA sequence is present in multiple copies in the genome, it will be present at higher concentrations, and this will allow reassociation to take place at low Co.t values. On the other hand, single DNA sequences will require a very much higher DNA concentration to reanneal, since being unique, the probability of collision between two of these complementary single-stranded chains is very low. When the degree of DNA reassociation of different eukaryotic organisms was measured on a wide range of Co.t values, it was possible to distinguish three different kinds of DNAs. One was reassociated at low Co.t values (10^{-4} to 10^{-1}), and was composed of highly repetitive sequences. A second was reassociated at intermediate Co.t values (10^0 to 10^2) and was composed of moderately repetitive or intermediate DNA sequences. The third was reassociated at even higher Co.t values, and corresponded to unique DNA sequences (Fig. 9.1). In man, for example, Arrighi and Saunders (1973) found that 35–40% of the genome was composed of repetitive DNA sequences, half of them highly repetitive, and the other half moderately repetitive or intermediate.

Compared to unique DNA sequences, repetitive DNAs are more difficult to understand in terms of function, organization, and evolutionary implications. Some classes of repetitive DNA sequences have known functions in man and other organisms, and their location in the human chromosome complement has been determined. This is the case of the 18S and 28S rDNA sequences (Henderson et al., 1972), and of the 5S rDNA sequences (Atwood et al., 1975), coding for ribosomal RNA. Other sequences of known function have been found to be repetitive in the eukaryotic genome, e.g., those coding for transfer RNA (Ritossa et al., 1966), histone messenger RNA (Kedes and Birnstiel, 1971), and mitocondrial DNA (Borst et al., 1967). However, such classes of repetitive DNAs do not add up to more than a percentage of the total DNA and have a relatively low repetition frequency.

Palindromes and Tandem Repeats

Repetitive DNA sequences occur as tandem repeats of various lengths (e.g., A.B.C.A.B.C. ...), or as reverse repeats, or palindromes (e.g., A.B.C.C.A.B.). Palindromes form hairpin duplexes as a result of the "folding-back" of a single-stranded DNA chain with a mutually inverted and adjacent repetitive sequence, even under conditions unfavourable for DNA reassociation, such as Co.t values below 10^{-4}. Hairpin duplexes have been observed under the electron microscope by Wilson and Thomas (1974). They have estimated that palindromes are composed of regions of approximately 300 1200 nucleotides long, arranged in clusters of two or four; their number in the human genome has been estimated at thousands or hundreds of thousands. They are widely distributed in the genome and their base composition does not differ significantly from the average base composition of the rest of the genome.

Tandem repeats may be of various length. It has been demonstrated that 80% of the human genome, following the *"Xenopus* pattern" (Davidson et al., 1975), is organized in single-copy sequences of approximately 2000 base pairs, which are interspersed with short repetitive sequences of approximately 400 base pair length (Schmid and Deininger, 1975). Moreover, the position of the short repetitive sequences was found to be random in relation to that of single-copy DNA sequences and inversely repetitive, thus suggesting that these three kinds of sequence were mutually interspersed. As against this, other tandem repeats consist of very large numbers of identical or closely similar short simple sequences that have been extensively amplified. This results in the very high amount of repetition of a simple unit characterized by few base pairs. As a consequence,

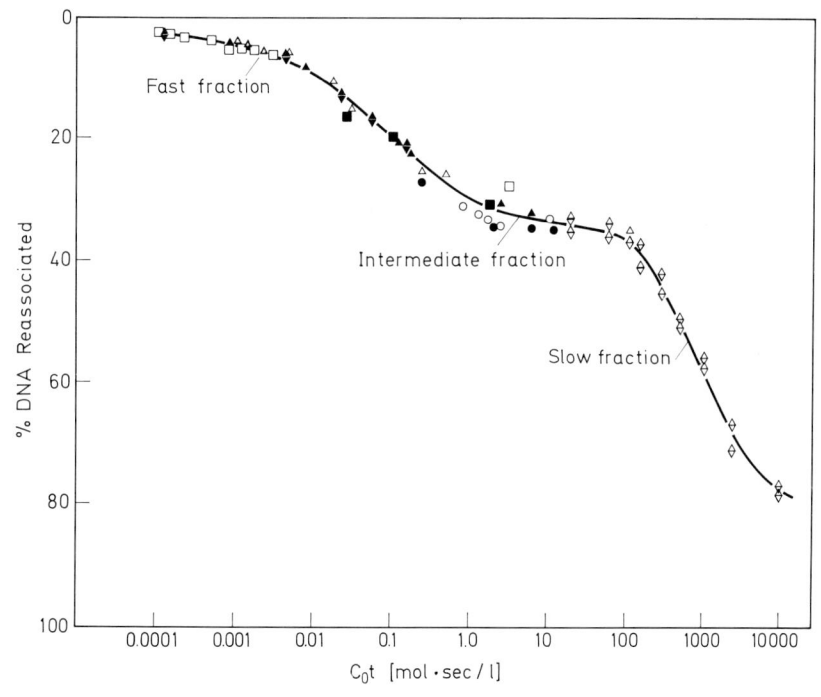

Fig. 9.1. Reassociation of human DNA was carried out in 0.12 M phosphate buffer (pH 6.8) at 60 °C. The reassociated fraction was taken as the DNA eluted from hydroxylapatite between 0.15 M and 0.3 M phosphate at 60 °C. □, △, ▲ represent HEp cell DNA at 1, 10, and 150 μg/ml. ◆, ■, ●, ○, ⬦ represent HEp cell DNA with placental DNA to a total concentration of 1, 10, 250, 730, and 12,343 μg/ml respectively. The result is expressed as a Cot curve. [Mitchell, Biochimica et Biophysica Acta *374*, 12–22 (1974)]

the base composition of these highly repetitive DNAs may differ from the average base composition of the genome. When this happens, it may be possible to detect and isolate them as separate satellite peaks of distinct buoyant density from the main bulk of the DNA, by centrifugation in caesium chloride. Light satellites will contain highly repetitive DNAs relatively richer in A-T; whereas heavy satellites will contain relatively higher amounts of G-C. When the difference is not marked or the highly repetitive DNAs do not differ significantly from the average base composition of the rest of the genome, separation is not possible by centrifugation in caesium chloride. Thus, separation of highly repetitive sequences has to be accomplished by other procedures. One of them makes use of the property of highly repetitive DNA to reassociate at low $C_0 \cdot t$ values, and of the property of hydroxylapatite to bind double-stranded DNA. This procedure, however, requires the shearing of the DNA into short-length segments, and does not allow the recovery and the physical separ-

Table 9.1. Cesium chloride densities of human satellite DNAs (g/ml)[a]

	Satellite DNA I	Satellite DNA II	Satellite DNA III	Satellite DNA IV
Native in neutral CsCl	1.687	1.693	1.696	1.700
Heat-denatured in 1 X SSC	1.703	1.704	1.715	1.716
Renatured at a $C_0t=0.1$	1.694	1.696	1.703	1.706
Separated strands in alkaline CsCl	1.707	1.740	1.740	1.730
	1.738	1.750	1.754	1.742
Density difference between complementary strands in alkaline CsCl	0.031	0.010	0.014	0.012
Separated strands in neutral CsCl	1.649	—	—	—
	1.712	—	—	—
Relative amount (%)	0.500	2	1.5	2
Position in Ag^+—Cs_2SO_4	Light	Heavy	Light	Light
Elution from a MAK column	Late	Early	Late	Early

[a] Source: Corneo et al. Satellite and repeated sequences in human DNA. In: Modern aspects of cytogenetics: constitutive heterochromatin in man, Symposia Medica Hoechst 6, pp. 29–37. Stuttgart: Schattauer Verlag, 1973.

ation of high molecular weight repetitive DNA sequences. For this reason, other methods are preferable, such as those that make use of the property of Ag^+ and Hg^{2+} ions to bind preferentially to G-C and A-T base pairs respectively, and using a preparative density gradient with Hg^{2+} Cs_2SO_4 and Ag^+ Cs_2SO_4. This method has been employed to isolate four major satellite DNAs in man by Corneo et al. (1967, 1968, 1970, 1971, 1972) since in man no distinct satellites can be observed and isolated from caesium chloride gradients. These four satellites[5] have been designated I, II, III, and IV, and each of them accounts for only a very small percentage of the total human genome, approximately 0.5, 2.0, 1.5, and 2.0% respectively. Altogether, they represent approximately 6% of the human genome, thus only a fraction of the total amount of highly repetitive DNA of man estimated by Arrighi and Saunders (1973). Satellite DNAs are characterized according to their density in neutral CsCl, in alkaline CsCl where strand separation can be analysed, and in neutral CsCl after denaturation and renaturation (Table 9.1). Their base composition can be measured analytically or can be indirectly estimated from their buoyant density (Schildkraut et al., 1962) or from their melting temperature (Marmur and Doty, 1962). It must be stressed that satellite II, although found on the heavy side of the main band DNA in Ag^+ Cs_2SO_4 gradient, is

---

[5] Other satellite DNAs have been isolated in man under different experimental conditions. For a review see Macaya et al., 1977.

a light satellite with a G–C-content of approximately 33.7% calculated from its buoyant density in neutral CsCl as against a G–C-content of 40% for total DNA (Mitchell, 1974). Satellite I is found on the heavy side of the main band in a Hg^{2+} Cs_2SO_4 gradient at pH 9.2 (Corneo et al., 1967, 1968), but on the light side in a $Ag^+Cs_2SO_4$ gradient at the same pH value (Corneo et al., 1971). Its G–C-content has been measured analytically and found to be approximately 26.4% (Schildkraut and Maio, 1969). Satellite III, also a light satellite, contains approximately 36% of G–C as estimated from its buoyant density and melting temperature (Macaya et al., 1977).

Satellite DNA and Sequence Heterogeneity

Analysis of satellite DNAs in mammals and other eukaryotes has shown that they may be composed of homogenous sequences as in *Drosophila* (Gall and Atherton, 1974) or of heterogeneous repeats containing a high amount of divergent sequences as in the guinea pig (Southern, 1970). Thus, highly repetitive DNAs may consist of related sequence families that have resulted from the amplification of a simple sequence, together with others that have diverged from the original basic sequence by mutation and nucleotide rearrangement. The degree of internal heterogeneity can be estimated by comparing the melting profiles of native and reassociated satellite DNA. Thermal denaturation of an homogeneous satellite DNA fraction (after reassociation) would occur in a steep line with little dispersion around its T_m value (the temperature at which half the DNA is denatured unter the conditions of the experiment). Thermal denaturation of a heterogeneous satellite fraction would, on the contrary, show a broader profile, and denaturation would occur over a wide range of temperatures around the value of T_m for reassociated DNA. Moreover, homogeneous satellite DNAs would form thermally stable double stranded molecules after reassociation, so that the T_m value of the reassociated DNA would be similar to or slightly below that of the native DNA. On the other hand, a satellite DNA in which there is a high amount of internal heterogeneity will contain, when reassociated, a considerable amount of mismatched bases, so that its thermal stability will be lower than that of the native satellite DNA. The difference (ΔT_m) between the T_m of the native and the reassociated satellite DNA would thus be indicative of the internal heterogeneity of the satellite fraction tested; it has been shown that lack of complementarity in 1.5% of the nucleotides will lower the thermal stability of a double stranded DNA molecule by 1 °C (Laird et al., 1969).

In man, intramolecular heterogeneity of three isolated satellite DNAs (I, II and III) showed different results. Satellite III showed the highest ΔT_m value (10 °C), whereas satellite II showed the lowest (2 °C), with

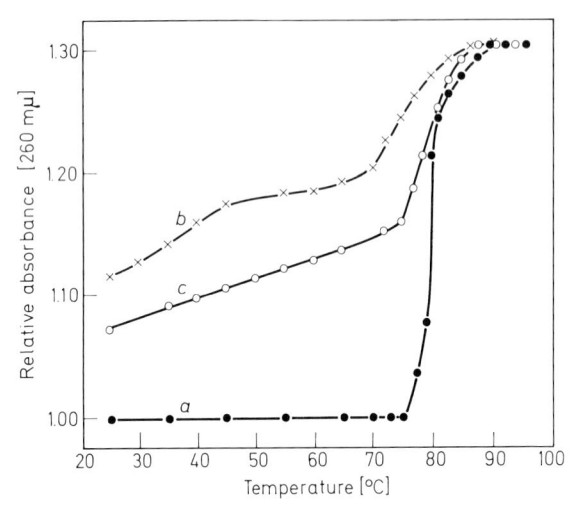

Fig. 9.2. Melting curves of isolated human satellite DNA I. *a* native; *b* previously heat denatured and fast cooled; and *c* previously denatured and renatured by hearing for 5 h at 65 °C in 2 × SSC. [Corneo et al., Journal of Molecular Biology 33, 331–335 (1968)]

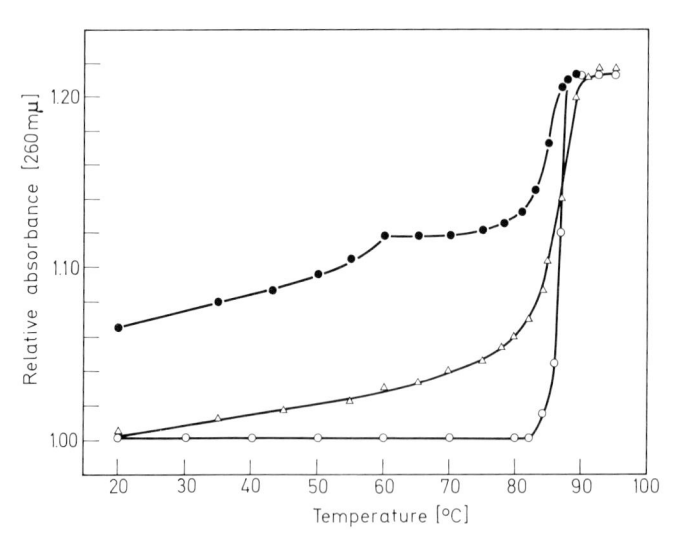

Fig. 9.3. Melting curves of isolated human satellite II DNA in 1 × SSC. (-○-○-) native; (-●-●-) previously heat-denatured and fast-cooled; and (-△-△-△-) heat denatured and renatured by heating at 65 °C for 5 h in 2 × SSC. [Corneo et al., Journal of Molecular Biology 48, 319–327 (1970)]

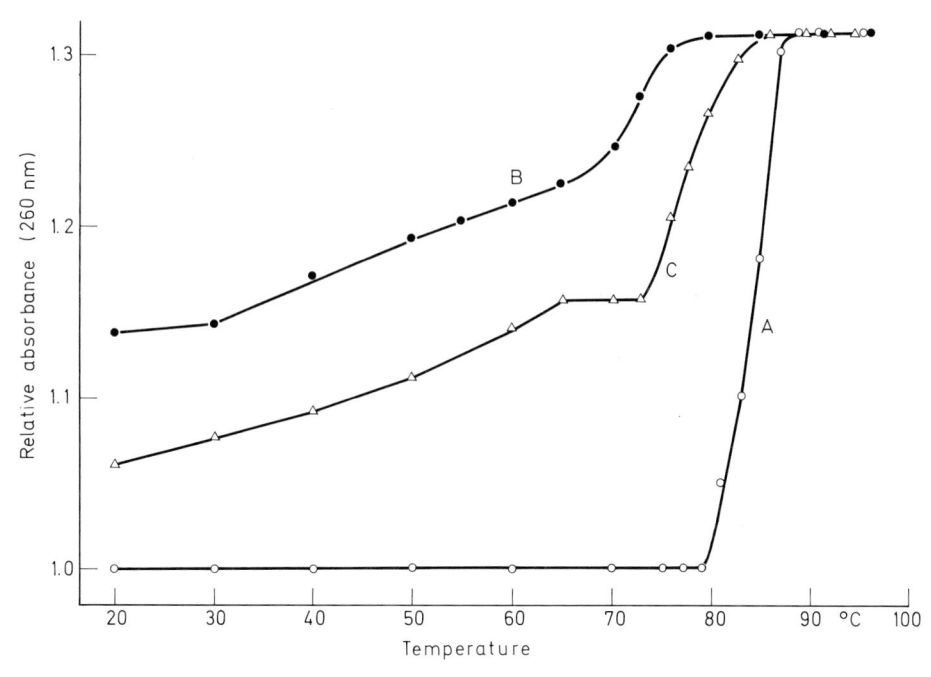

Fig. 9.4. Melting curves of isolated human satellite III DNA in 1 × SSC. *A* native; *B* previously heat denatured and fast cooled; *C* heat-denatured and then renatured by heating at 60 °C for 20 h in 2 × SSC. [Corneo et al., Biochimica et Biophysica Acta *247*, 528–534 (1971)]

an intermediate value for satellite I (6 °C; Figs. 9.2, 9.3, and 9.4). These results suggest that the three human satellites have undergone different amounts of evolutionary divergence, as remarked by Jones (1977), a point that will be extensively discussed in the chapters to come. However, since evolution has resulted in change in both the non-repetitive and the repetitive DNA sequences of the human and primate genome, the three following chapters will first deal with evolutionary change at the non-repetitive DNA level; another at the protein level; and the third, where structural gene sequences have been localized by methods of gene mapping, at the chromosome level. In four subsequent chapters, evolution of repetitive DNA sequences in the primate genome will be discussed at the DNA level, and in relation to their chromosome distribution.

References

Arrighi, F. E., Saunders, G. F.: The relationship between repetitious DNA and constitutive heterochromatin with special reference to man. In: Modern aspects of cytogenetics: constitutive heterochromatin in man. Pfeiffer, R. A. (ed.), Vol. 6, pp. 113–133. Stuttgart, New York: Schattauer Verlag 1973

Atwood, K.C., Yu, M.T., Johnson, L.D., Henderson, A.S.: The site of 5S RNA genes in human chromosome 1. Cytogenet. Cell Genet. *15*, 50–54 (1975)

Borst, P., Ruttenberg, G., Kroon, A.M.: Mitochondrial DNA. I. Preparations and properties of mitochondrial DNA from chick liver. Biochim. Biophys. Acta *149*, 140–155 (1967)

Britten, R.J., Kohne, D.E.: Repeated sequences in DNA. Science *161*, 529–540 (1968)

Corneo, G., Ginelli, E., Polli, E.: A satellite DNA isolated from human tissues. J. Mol. Biol. *23*, 619–622 (1967)

Corneo, G., Ginelli, E., Polli, E.: Isolation of the complementary strands of a human satellite DNA. J. Mol. Biol. *33*, 331–335 (1968)

Corneo, G., Ginelli, E., Polli, E.: Repeated sequences in human DNA. J. Mol. Biol. *48*, 319–327 (1970)

Corneo, G., Ginelli, E., Polli, E.: Renaturation properties and localization in heterochromatin of human satellite DNAs. Biochim. Biophys. Acta *247*, 528–534 (1971)

Corneo, G., Ginelli, E., Zardi, L.: Satellite and repeated sequences in human DNA. In: Modern aspects of cytogenetics: constitutive heterochromatin in man. Symposia Medica Hoechst. Pfeiffer, R.A. (ed.), Vol. 6, pp. 29–37. Stuttgart: Shattauer-Verlag 1973

Corneo, G., Zardi, L., Polli, E.: Elution of human satellite DNAs on a methylated albumin kieselguhr chromatographic column. Isolation of Satellite IV DNA. Biochim. Biophys. Acta *269*, 201–204 (1972)

Davidson, E.H., Galau, G.A., Angerer, R.C., Britten, R.J.: Comparative aspects of DNA organization in metazoa. Chromosoma (Berlin) *51*, 253–259 (1975)

Gall, J.G., Atherton, D.D.: Satellite DNA sequences in *Drosophila viridis*. J. Mol. Biol. *85*, 633–664 (1974)

Henderson, A.S., Warburton, D., Atwood, K.C.: Location of ribosomal DNA in the human chromosome complement. Proc. Nat. Acad. Sci. USA *69*, 3394–3398 (1972)

Jones, K.W.: Repetitive DNA and primate evolution. In: Molecular structure of human chromosomes. Yunis, J.J. (ed.), pp. 295–326. New York: Academic Press 1977

Kedes, L.H., Birnstiel, M.L.: Reiteration and clustering of DNA sequences complementary to histone messenger RNA. Nature New Biol. (London) *230*, 165–169 (1971)

Laird, C.L., McConaughy, McCarthy, B.J.: Rate of fixation of nucleotide substitutions in evolution. Nature (London) *244*, 149–154 (1969)

Macaya, G., Thiery, J.P., Bernardi, G.: DNA sequences in man. In: Molecular structure of human chromosomes. Ed. Yunis, J.J. (ed.), pp. 35–58. New York: Academic Press 1977

Marmur, J., Doty, P.: Determination of the base composition of deoxyribonucleic acid from its thermal denaturation temperature. J. Mol. Biol. *5*, 109–118 (1962)

Mitchell, A.R.: Properties of the homogeneous main band DNA from the human genome. Biochim. Biophys. Acta *374*, 12–22 (1974)

Ritossa, F., Atwood, K.C., Lindsley, D.L., Spiegelman, S.: On the chromosomal distribution of DNA complementary to ribosomal and soluble RNA. Nat. Cancer Inst. Monogr. *23*, 449–472 (1966)

Schildkraut, C.L., Maio, J.J.: Fractions of HeLa DNA differing in their content of guanine plus cytosine. J. Mol. Biol. *46*, 305–312 (1969)

Schildkraut, C.L., Marmur, J., Doty, P.: Determination of the base composition of deoxyribonucleic acid from its buoyant density in CsCl. J. Mol. Biol. *4*, 430–443 (1962)

Schmid, C.W., Deininger, P.L.: Sequence organization of the human genome. Cell *6*, 345–358 (1975)

Southern, E.M.: Base sequence and evolution of guinea-pig α-satellite DNA. Nature (London) *227*, 794–798 (1970)

Wilson, D.A., Thomas, C.A.: Palindromes in chromosomes. J. Mol. Biol. *84*, 115–144 (1974)

Chapter 10 Evolution of Non-Repetitive DNA Sequences in Man and the Great Apes

Nucleotide Substitutions and Phyletic Divergence

The main reason why species have diverged from each other is because mutant substitutions have accumulated and become fixed in their genomes. This has inevitably resulted in classes of DNA which are unique for each species, as well as DNA sequences which, although shared by many species, have accumulated different mutant substitutions. In this respect, a comparison of the DNA level between phylogenetically related species is a powerful tool to estimate the amount of evolutionary change which has taken place between them. These studies are based on three different properties of DNA. One is its property to denature and reaneal under experimental conditions and the possibility of separating double-stranded DNA from single-stranded DNA by hydroxylapatite binding (Britten and Kohne, 1968). Another is the possibility of forming double-stranded (heteroduplex) DNA by incubating single-stranded DNA of different species, provided there is sufficient complementarity between the DNA sequences of the species tested. A third important property is that heteroduplex DNA molecules show a lower thermal stability than reassociated DNA molecules of each of the two individual species tested (homoduplex), as a consequence of the higher amount of mismatching which inevitably occurs in heteroduplexes. Mismatching results from lack of complementarity between the DNA of the species tested, which in turn results from the number of nucleotide substitutions or replacements which have occurred in the DNA of each species after they diverged from the common ancestor. Laird et al. (1969) have estimated that lack of complementarity in 1.5% of the nucleotides will lower the thermal stability of double-stranded DNA molecules by 1 °C, and this formula is used to estimate the number of nucleotide substitutions which have occurred in the DNA of different organisms. Moreover, if these data are related to the time elapsed since the evolutionary divergence of the species under test, it is possible to estimate the rate of nucleotide substitution in a lineage.

This approach is, however, restricted to the study of non-repetitive DNA sequences, since repetitive DNAs comprise families of similar (but not identical) sequences whose member sequences may reassociate together. Although phylogenetically related species may contain the same families as a result of their being present in a common ancestor, the degree of similarity between these families at the time of divergence is unknown, and this renders impossible the determination of the rate of nucleotide substitution. Non-repetitive DNA sequences, on the other hand, comprise sequences which are unique per haploid genome. Reassociated non-repetitive DNA shows a high thermal stability as a consequence of a high degree of complementarity or matching. If a non-repetitive DNA sequence is found in two or more phylogenetically related species, it must have derived from an ancestral sequence carried by a recent common ancestor, and the rate of nucleotide substitution can be precisely estimated. Kohne et al. (1972) used this approach to study the evolution of primate non-repetitive DNA sequences, and estimated the rate of nucleotide replacement occurring during the divergence of seven primates, including man. The degree of divergence between species, e.g., man and chimpanzee, was estimated by the difference in the T_m (ΔT_m) between human non-repetitive reassociated DNA (human homoduplex) and human-chimpanzee heteroduplex DNA. This study indicated that man and chimpanzee only differed in 2.4% of their nucleotides as a result of substitutions which had taken place since their divergence as different species, approximately 15×10^6 years ago. Similar studies were carried out with heteroduplexes formed between human DNA and DNA of other primate species. The T_m of each heteroduplex was subtracted from that of human homoduplex, so that this latter was used as an indicator of relatedness, and results indicated how similar each of the DNAs tested was from human DNA. It was clear that closely related species showed smaller amounts of nucleotide substitution. As an example in the Hominoidea, the results of Kohne et al. (1972) showed that 5.3% nucleotide replacements had occurred between man and the gibbon, more than twice that existing between man and the chimpanzee. A second set of experiments used heteroduplex DNAs between green monkey DNA and other primate DNAs, and their T_m values were substracted from the T_m of reassociated non-repetitive DNA of the green monkey (green monkey homoduplexes). In this experiment, green monkey homoduplex DNA was the indicator of relatedness, and results indicated how similar each of the DNAs tested was from green monkey DNA. By combining the values of ΔT_m obtained with the two indicators of relatedness (human and green monkey homoduplexes) Kohne et al. (1972) constructed a phylogenetic tree and estimated the amount of change or percentage of nucleotide substitution in different periods of primate evolution using different assumptions. For example, the period encompassed between the branching off of the gibbon from the common stock of the Hominoidea and the splitting of the lineage of man and chimpanzee (15×10^6 years) has accounted for 1.5% substitutions, or approximately 0.1% per 10^6 years. The period following the splitting of man and chimpanzee (15×10^6 years) has probably accounted for 1.2% substitutions in each lineage (assuming that the 2.4% observed between man and chimpanzee accounts for half the amount in each lineage), thus approximately 0.08% per 10^6 years in each lineage. When the percentages of nucleotide replacement between different periods of primate evolution were compared, and when the rates of substitution were estimated in terms of absolute time, Kohne et al. (1972) found that nucleotide substitution had been faster at early stages of primate evolution than during later stages.

The main factor involved in the slowing up of nucleotide substitution was thought to be the lengthening of generation time, which in man is considerably longer than in monkeys.

Man and the Great Apes: Phylogenetic Implications

Experiments in which heteroduplex DNA molecules were made between human and great ape DNA have been informative in determining man's position in the classification. Hoyer et al. (1972) carried out experiments with non-repetitive DNA on man, chimpanzee, gorilla, orangutan, gibbon and green monkey, and used the human and the orangutan homoduplex DNAs as alternative criteria of relatedness. When human homoduplex DNA was used as an indicator of relatedness, the ΔT_m value of the human-orangutan heteroduplexes was twice that of the human-gorilla, and four times that of the human-chimpanzee heteroduplexes. On the other hand, when orangutan homoduplex DNA was used as an indicator of relatedness, the ΔT_m values of the orangutan-chimpanzee and of the orangutan-gorilla heteroduplexes were greater than the previously found ΔT_m values of the human-chimpanzee and the human-gorilla heteroduplexes respectively. Of these two experiments, the first showed that orangutan DNA was less complementary to human DNA than was the DNA of the two African apes, and the second experiment showed that the DNA of each of the African apes was less complementary to the orangutan DNA than each of them had previously been to human DNA. This proved that the orangutan was further away from man than were the African apes, but could not determine, however, which species of African ape was closer to man, since neither chimpanzee or gorilla homoduplex DNA was used as an indicator of relatedness. More recently, Beneviste and Todaro (1976) have solved this problem by carrying out similar experiments in which human, chimpanzee, and gorilla homoduplexes were used as indicators of relatedness to the heteroduplexes formed between any two of them. The results of Beneviste and Todaro (1976) showed that man, chimpanzee, and gorilla were equally distant from each other. A phylogenetic tree of the Homininae as proposed by Goodman (1975) can thus be represented as a trichotomy, each of the three terminal points being equidistant from each other.

Is Man an Asian Ape?

In Chap. 2 we have discussed the palaeontological evidence in favour of and against the Asian origin of man. It is interesting now to reconsider that problem from the evidence supplied by experiments of DNA hybridisation as reported by Beneviste and Todaro (1976). These experiments are based on the fact that the genome of all old-world primates (African and Asian monkeys, apes and man) possess a gene sequence related to the RNA of a C-type virus isolated from a group of African monkeys, the baboons (e.g., *Papio cynocephalus*). A baboon-type C viral DNA probe was hybridised to the non-repetitive DNA of different species of old-world primates; the thermal stability of each kind of hybrid DNA molecule was compared to the thermal stability of the viral C DNA-*Papio cynocephalus* DNA, used as indicator of relatedness. The ΔT_m thus estimated was found to be a function of the phylogenetic distance of each species tested from the baboon; those closely related to it (e.g., other baboon species) showed low ΔT_m values, and those more distant (e.g., chimpanzee and gorilla) showed higher ΔT_m values. However, the decrease in thermal stability was not always positively correlated to the phylogenetic distance of the species tested from the baboon, but rather related to their geographic distribution. A good example of such findings comes from the comparison of the ΔT_m obtained from C-DNA hybrids with DNA of macaques (Asian monkeys) and with DNA of mangabeys (African monkeys). The macaques (e.g., *Macaca mulata*) and the mangabeys (e.g., *Cercocebus atys*) are approximately equally distant from the baboon. This was proved by measuring the ΔT_m of heteroduplex DNA molecules of each of these two species with baboon DNA, using baboon homoduplex

DNA as an indicator of relatedness, and obtaining practically identical values of ΔT_m. However, the ΔT_m values of the hybrids C-DNA-macaque DNA and the C-DNA-mangabey DNA were clearly different, thus indicating that the virogene sequences of the two species were quite distinct. Those of the Asian species (macaques) produced higher values of ΔT_m than those of the African species (mangabeys), thus showing that Asian species have undergone more nucleotide substitutions in the same virogene sequence. When similar experiments were carried out with the ape and human DNA, it was shown that the thermal stability of C-DNA-human DNA hybrid was very low, as were those of the hybrid molecules formed between C-DNA and DNA of *Hylobates, Symphalangus* and *Pongo,* all of them species of Asian origin. On the other hand, the virogene sequence in two species of African apes (chimpanzee and gorilla) showed closer similarities to that of the baboon, since the ΔT_m values of the hybrid molecules formed between C-DNA and DNA of each of these two species was considerably lower than those found for human and Asian ape DNA-C DNA hybrids. The important conclusion of this experiment was that the virogene sequence of man appeared to have undergone an amount of nucleotide substitution similar to that found in Asian apes, whereas the virogene sequence in African apes was closer to that found in African monkeys. Beneviste and Todaro (1976) have commented on the fact that the type of C virogenes among anthropoid primates and their evolutionary persistence for 30–40 million years may be an indication that species possessing them have a selective advantage. One of them could be the resistance to C viral infection, since endogenous C viral genes restrict the replication of highly related viruses in the same cells, a function for which African primates appear to be under higher selective pressure than Asian primates. The fact that the virogene sequence has undergone substantial nucleotide substitution in man is then an indication that man's direct ancestors must have evolved in Asia, where the selective pressure for the virogene sequence was low probably during all the Pliocene; and that the appearance of hominids in Africa may have resulted from extensive migration.

References

Beneviste, R.E., Todaro, G.J.: Evolution of type C viral genes: evidence for an Asian origin of man. Nature (London) *261*, 107–108 (1976)
Britten, R.J., Kohne, D.E.: Repeated sequences in DNA. Science *161*, 529–540 (1968)

Goodman, M.: Protein sequence and immunological specificity. Their role in phylogenetic studies of the primates. In: Phylogeny of the primates. Luckett, W.P., Szalay, J.S. (eds.), pp. 219–248. New York: Plenum Press 1975
Hoyer, B.H., van de Velde, N.W., Goodman, M., Roberts, R.B.: Examination of hominid evolution by DNA sequence homology. J. Hum. Evol. *1*, 645–649 (1972)
Kohne, D.E., Chiscon, J.A., Hoyer, B.H.: Evolution of primate DNA sequences. J. Hum. Evol. *1*, 627–644 (1972)
Laird, C.L., McConaughy, McCarthy, B.J.: Rate of fixation of nucleotide substitutions in evolution. Nature (London) *244*, 149–154 (1969)

Chapter 11 Evolution of Structural Gene Sequences

Missense Mutations and Amino Acid Substitutions

A nucleotide substitution would result in an amino acid substitution only if a missense mutation were produced in a DNA sequence which is transcribed and translated. Nonsense mutations could produce only shorter polypeptides by premature chain termination, while samesense mutations will be unnoticed, unless they modify the rate of translation of the messenger RNA transcribed from the mutated cistron. Frame-shift mutations (which are due to deletions or insertions of nucleotides that do not result in deletions or insertions of entire codons) could result in totally different proteins. Most probably, a frame-shift mutation would not be tolerated, but if it were, it would be most difficult to trace by any of the biochemical methods used in protein analysis. This is because a frame-shift mutation may result in an amino acid sequence entirely different from that of the original protein until the two of them would have practically nothing in common. Thus, they would lack the minimum degree of "homology", or structural and functional similarity, necessary to make a comparison between the original and the mutated protein. Therefore, unless a frame-shift mutation is somehow traced at the DNA level, it is unlikely to be detected by comparing proteins of different organisms and assigning them to the same structural cistron. Excluding tRNA mutations that may cause ambiguous coding, the amount of amino acid substitutions between homologous proteins of different species is therefore indicative of the amount of missense mutations that they have undergone at their structural cistrons. However, missense mutations represent only a fraction of the overall number of mutations produced by nucleotide substitution. Whitfield et al. (1966) have estimated that if all possible kinds of nucleotide substitutions occurred at random in the genetic code, the relative frequency of nonsense: missense: samesense mutations would be 1:17:6.

Up to the present, extensive studies have been carried out in comparing homologous proteins of different primate species including man (Goodman, 1976), and a detailed analysis of this problem would inevitably go beyond this work. However, it is important to mention some of the most important implications of this study of proteins in the general context of human evolution and particularly in relation to chromosomes. A comparison at the molecular level has led scientists to construct phylogenetic trees under different assumptions. One of them assumes that the more ancient the common ancestor of a pair of species, the greater will be the genetic distance between them (divergence hypothesis). A divergence tree based on the degree of similarity between species is then constructed, those showing the smallest distance being considered the closest related. This is the approach of the phenetic school (Sokal and Michener, 1958), and as we have observed in Chap. 7, it may not necessarily hold good if the evolutionary rate of change has not been uniform, or if similarity is a result of parallel evolution or convergence. A second approach is to estimate the degree of divergence between species within an additive context in a clade. This is done by comparing a clade of species to a more distant species used as reference. This latter is needed to estimate how much change has taken place along each lineage of the clade, or in other words, the outside reference is needed to apportion the minimum mutation distances into sets of lengths connecting each exterior point in a cladogram. Take for example a clade of species A, B, C, D, and a fifth species (E), more distant from them, used as reference. Suppose that the number of observed differences between any pair of species is that of the matrix in Fig. 11.1. In the first instance, we can draw a cladogram grouping species, but we ignore a priori how many of these differences have occurred along or within each lineage. Say for example that species A and B show 14 mutant substitutions when an homologous protein is compared. How can we apportion the line which goes from A to B, thus knowing how many changes took place in each lineage after the two species diverged? In theory 14 substitutions between two species could result from either $1+13$; $2+12$; $3+11$ ... etc. changes occurring in each lineage. However, a cladistic analysis of additive data can answer this question as illustrated in Figure 11.1, except that the line connecting the outside species (E) to the clade cannot be apportioned.

The importance of this method is that it allows the estimation of relative amounts of change inside a clade independently of the time of evolutionary divergence. In this way it eliminates uncertainties which result from imprecise estimations of evolutionary time spans. This hypothesis, however, suffers from the tendency to underestimate minimum mutation distances between anciently separated species more grossly than between recently separated ones. This is due to the fact that the longer the evolutionary span, the higher the probability of a codon to suffer

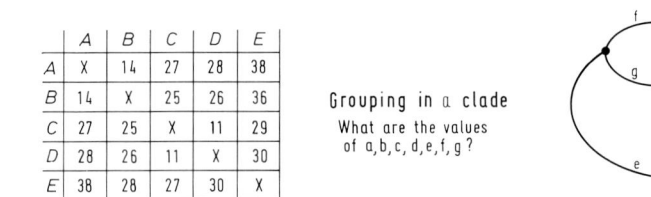

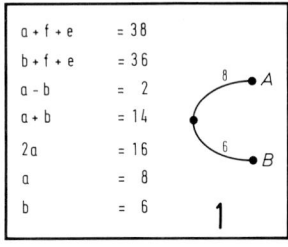

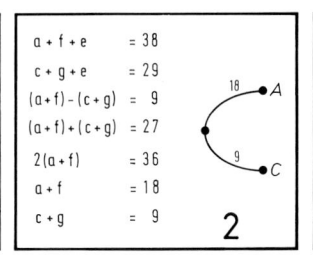

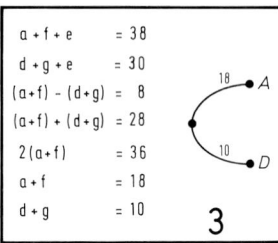

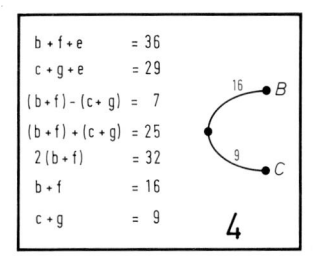

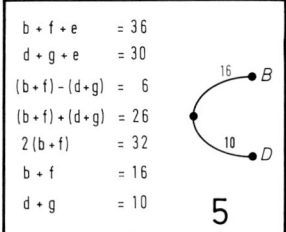

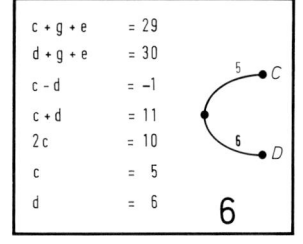

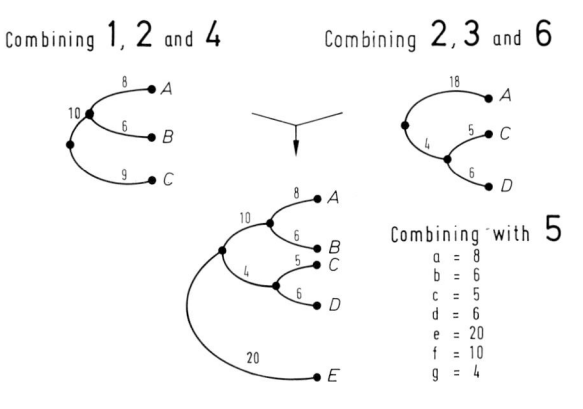

Fig. 11.1. Construction of a cladogram following an additive hypothesis. Units of distance (e.g., amino acid substitutions) between species A, B, C, D and E

more than one missense mutation. If this happened, alternative possibilities could occur: one, that a second missense mutation reverts to the original codon; the other, that an entirely new codon, different from any of the previously existing ones, is formed. The first possibility (back mutation) would pass unnoticed when comparing two homologous proteins and would be the same as if no nucleotide substitution had even taken place at that codon. The second will be reckoned as one missense mutation instead of two. Since the occurrence of these situations may not necessarily be evenly distributed in all lines of descent, the additive approach may lead to less precise results than expected. Nonetheless, it is important to consider some of its relevant findings concerning the origin of man before examining the third approach, based on the principle of maximum parsimony.

Molecular Evolutionary Clocks and the Human-Ape Divergence

When the number of albumin plus transferrin units of change (estimated by immunological distances) along eight primate lineages, including the Hominoidea, were compared in an additive tree, it was found that they fell into a narrow range (Sarich and Cronin, 1976).The average number corresponded to 137.5 units with a S.D. = 8.569. Findings of this kind led to the postulate that rates of amino acid substitution of homologous proteins (e.g., albumins) had remained constant during different lines of descent. Thus, the process of amino acid substitution as observed in albumins and transferrins could be envisaged as a molecular clock, each tick of the clock being marked by the fixation of a new mutation. If the rate of fixation of new mutations were constant, the amount of change that one lineage has undergone in relation to another would be indicative of their relative age. For example, Sarich and Cronin (1976) found approximately equal numbers of changes in albumin and transferrin within the Hominoidea and the Cercopithecoidea, thus indicating that both lineages were equally old. The same calculations showed that about twice the amount of change had occurred within the Platyrrhini as within the Hominoidea or the Cercopithecoidea, thus indicating that the branching off of the Platyrrhini from the Anthropoidea was twice as old as the Hominoid–Cercopithecoid divergence. However, since this method only showed the relative age of each lineage, the finding of absolute ages required the "calibration" of the molecular clock at one point of divergence in the phylogenetic tree where the fossil record showed well-documented evidence. Sarich and Cronin (1976) reckoned the beginning of the primate radiation to have taken place 70 million years ago, thus estimating further back the marsupial–placental divergence at 125 million years, and more recently, the catarrhine–platyrrhine divergence at 35 million years. The extrapolation of such estimates indicated, however, that the branching off of the Cercopithecoidea from the Hominoidea occurred at approximately 20 million years, that of the gibbons and orangutan from the Hominoidea stock at about 10 million years, and the divergence of man, chimpanzee, and gorilla only at 4 million years. This calculation contradicts the present evidence supplied by the fossil record in which our first presumed ancestor, *Ramapithecus,* probably diverged from the common Hominoid stock about 15 million years ago (Chap. 2).

A further implication from the existence of molecular evolutionary clocks is that we might explain evolution as resulting from the fixation of a greater number of selectively neutral mutations contrary to the Darwinian conception of selective advantage. If we envisaged each lineage in the phylogeny as a divergent branch of a tree, each of the branches composed of organisms occupying different environments, undergoing adaptive radiation, and suffering different fluctuations in population size, we would expect each lineage to be subjected to a different selective pressure. For this reason, if mutant substitutions were fixed as a consequence of conferring selective advantage, we would not expect fo find molecular evolutionary clocks, or constant mutation rates. If they existed, this would imply that independently of the selective pressure which evolving organisms were bound to bear, they were capable of fixing similar numbers of substitutions per similar evolutionary spans. Although this would not prove that the great majority of mutations were actually neutral, a neutral mutation theory necessarily implies and needs the existence of molecular clocks. This theory, initially put forward by Kimura (1968, 1976) and by King and Jukes (1969) is based on the fact that the average rate of amino acid substitution estimated by comparing homologous proteins of different organisms was approximately 1.7×10^{-9} per amino acid site (Kimura, 1976). When extrapolated to the total number of nucleotide sites of the mammalian genome, this rate became equal to 2.5 per year, or approximately 2.5×3 per generation taking 3 years as the average generation time of mammals. Since the mutation rate per generation thus obtained was more than 2000 times the number of mutant substitutions that organisms could bear without being exterminated by selection (Haldane, 1957; see Chap. 4), Kimura (1968) proposed that the great majority of amino acid replacements resulted from the random fixation of selectively neutral mutations. This theory has been criticised on two grounds. One is that the extrapolation of the average rate of amino acid substitution to the total number of nucleotide sites of the mammalian genome overlooked the fact that structural genes comprise only a small percentage of the eukaryote genome in which considerable amounts of DNA were found to be repetitive and apparently meaningless. The second, as pointed out in Chap. 4, is that Haldane's estimates are based on the assumption of multiplicative fitness, and that on the assumption of threshold fitness the observed and the expected mutation rates would be compatible with the theory of gene fixation by natural selection.

The Maximum Parsimony Approach and the Decelerated Rates of Molecular Evolution in Higher Primates and Man

The maximum parsimony procedure reconstructs a phylogeny by distinguishing patristic and convergent similarities from derived similarities which appeared in a more recent common ancestor. These, in turn, are used to determine the cladistic relationship of the species in the phylogenetic tree. When applied to molecular evolution, this method reconstructs a phylogenetic tree by the minimum amount of amino acid substitutions needed to explain the differences between homologous proteins. This approach has the advantage of maximizing the number of genetic similarities resulting from common ancestry, and is less biased by the occurrence of back and parallel mutations than other approaches (Goodman et al., 1972). Its main disadvantage is that differences between species must be referred to time spans of evolutionary divergence estimated from the fossil records which are not always precise or available. However, in spite of its limitations this approach has supplied a more coherent view of molecular evolution than any other. In the first place, man and the great apes have been found to form a cluster of very similar species, a condition which, contrary to previous classifications, has led Goodman (1975) to divide the Hominoidea into Hylobatidae (gibbons) and Hominidae (man and the great apes; see Chap. 3). The second important conclusion was that the rates of amino acid substitution observed in a number of proteins (α- and β-globins, myoglobin, carbonic anhydrase, and cytochrome C) were variable during phylogeny (Goodman, 1976). Although approximately constant rates were found during the period of eutherian radiation including that of the lower primates, the higher primates, among which man is included, showed a marked deceleration in the rate of nucleotide replacements. Such deceleration must be responsible for the remarkable similarity at the biochemical level that is observed between man and the chimpanzee, for example, as pointed out by King and Wilson (1975), if we consider the ape–man divergence as having taken place at approximately the 15 million years indicated by the fossil record. In other cases, mutant substitutions appear to have occurred at fast rates during phylogeny. An example of such acceleration is that of the rate observed in carbonic anhydrase II in the period between the appearance of the common eutherian ancestor and that of the common catarrhine ancestor (Tashian et al., 1976), to be later followed by a profound deceleration during catarrhine radiation.

Whence Come Chromosomes?

Although biologists have now a better idea of how genes have evolved, the question of how entire organisms have evolved is still far from being understood. Goodman (1976) has attempted to interrelate molecular and organic evolution, suggesting that accelerations and decelerations in mutation rates occurred in periods of different organic adaptations. In his view, molecular evolution occurred as a trial and error process in which high mutation rates occurred in periods of early phyletic divergence, affecting proteins that had not yet acquired their maximum degree of potential perfection. Those mutant substitutions which were fixed must have conferred a higher degree of "functional density"[6] on the protein in which they occurred. As a consequence, increased protein complexity resulted in a higher degree of internal integration by which organisms became more independent from the fluctuations of their external environments. Furthermore, evolution must have become more conservative if organisms were to maintain their highly perfected internal integration once it had been achieved, so that high mutation rates would subsequently not have been tolerated. In Goodman's opinion, any important leap forward in evolution must have been preceded by a considerable amount of molecular change to be later followed by a decelerated mutation rate, later radiations suffering less amount of change than earlier ones as a result of having inherited a larger number of successful adaptations.

On the other side, Wilson et al. (1974a) have shown that molecular and organic evolution have followed opposite directions. Frogs, for example,

---

[6] Functional density is defined as the proportion of sites in a protein involved in specific functions.

which comprise an organically homogeneous group, have undergone considerably larger amounts of molecular change in albumin than mammals, although mammals comprise a group of extreme organic adaptations. Moreover, the mean albumin difference between pairs of mammals and frog species capable of producing heterospecific offspring, showed that while in mammals it accounted for only 3 units, or approximately 0.6% sequence difference between species, it was 36 between frogs, or approximately 12 times higher than between mammals. These findings indicated that molecular evolution and organic evolution were unrelated. The latter must have resulted from change in non-structural genes, probably in regulatory genes. When molecular change was compared to chromosome change in both mammals and frogs, it was shown that mammals differing in only 6 albumin units had a 50% chance of having a different chromosome number, whereas in frogs the same amount of molecular change occurred between species of identical chromosome number (Wilson et al., 1974b). It was also found that for frogs to have a 50% probability of differing in chromosome number, their albumin difference had to be equal to 120 units, or approximately 20 times higher than mammals. Based on previous estimates that the rate of mutant substitutions in albumin was equal in mammals and frogs (1.7 units/10^6 years), the rate of change in chromosome number in mammals was estimated as 1 every $6/1.7 = 3.5 \times 10^6$ years, whereas that of frogs only 1 every $120/1.7 = 70 \times 10^6$ years. However, the estimate that chromosome change has occurred 20 times faster in mammals than in frogs would not necessarily hold if the rate of mutant substitution of protein were demonstrated to occur at a higher rate in frogs than in mammals. In fact, Goodman (1976) has found in a maximum parsimony tree of β-globin that the differences observed between two species of the genus *Rana* (spp. *esculenta* and *catesbiana*) were larger than between any two orders of placental mammals. When referred to the time scale elapsed since the divergence of *Rana,* and that of placental mammals, the mutation rate in these two frogs accounted for 0.72 nucleotide replacements/10^6 years as against 0.29 in mammals, or approximately 2.4 times faster. Then, if similar differences existed for albumin rates as for β-globins between frogs and mammals, chromosome change in mammals could have taken place every $6/0.29 = 21 \times 10^6$ years as against $120/0.72 = 167 \times 10^6$ years in frogs, thus making the rate of chromosome change in mammals only eight times faster than that of frogs.

In order to avoid imprecise estimations of chromosome change from protein mutation rates, Wilson et al. (1975) analysed different taxonomic categories (species, genus, family, order, and class) of vertebrates and molluscs in relation to the chromosome change observed within each of them. These estimates showed that chromosome change increased from the first group (intraspecific), which showed a mean value close to zero, to the fifth group (intraclass), which was approximately 100%. Furthermore, chromosome change observed between species of the same genera was found to be uncorrelated to the time elapsed since the evolutionary divergence of the species tested. Findings of this kind suggested that organic and chromosomal evolution had been positively correlated, a finding that coincided with the observation that placental mammals had undergone chromosome change at a rate five times higher than other vertebrates and molluscs which comprise groups where organic evolution has been more conservative. This correlation, although general, is less than universal when applied to individual cases. For example, the karyotype of the dolphin *(Delphinus dolphin)* and that of the Sei whale *(Baleanoptera borealis)* have the same diploid number (44), and are morphologically similar, except for one chromosome pair of small submetacentric chromosomes in the dolphin that has been substituted by a pair of large subtelocentric chromosomes in the Sei whale (Arnason, 1972). In fact Mysticetes, or baleen whales, showed a diploid chromosome number (44) identical to most Odontocetes, or toothed whales, notwithstanding that these two groups belong to different suborders which had even been considered to be of diphyletic origin. On the other side, two species belonging to the same genus *(Muntiacus)* such as the chinese muntjack (sp. *reevesi*), and the indian muntjack (sp. *muntjack*) showed completely different karyotypes; the former with a diploid number = 46, and the latter with $2n = 6♀, 7♂$ (Wurster and Benirschke, 1970). Finally, the relative conservation of human and great ape chromosomes does not explain why and how man has achieved his high degree of organic complexity compared to other primates. However, now that we have analysed the evolution of non-repetitive sequences at the DNA level and at the protein level, let us see how our structural genes have been transmitted in chromosomes for the past 35 million years.

References

Arnason, U.: The role of chromosomal rearrangement in mammalian speciation with special reference to Cetacea and Pinnipedia. Hereditas 70, 113–118 (1972)

Goodman, M.: Protein sequence and immunological specificity. Their role in phylogenetic studies of the primates. In: Phylogeny of the primates. Luckett, W. P., Szalay, J. S. (eds.), pp. 219–248. New York: Plenum Press 1975

Goodman, M.: Towards a genealogical description of the primates. In: Molecular anthropology. Genes and proteins in the evolutionary ascent of the primates. Goodman, M., Tashian, R. E., Tashian, J. (eds.), pp. 321–353. New York, London: Plenum Press 1976

Goodman, M., Barnabas, J., Moore, G. W.: Man, the conservative and revolutionary mammal. Molecular findings of this paradox. J. Hum. Evol. *1*, 663–685 (1972)

Haldane, J. B. S.: The cost of natural selection. J. Genet. *55*, 511–524 (1957)

Kimura, M.: Evolutionary rate at the molecular level. Nature (London) *217*, 624–626 (1968)

Kimura, M.: How genes evolve; a population geneticist's view. Ann. Genet. *19*, 153–168 (1976)

King, J. L., Jukes, T. H.: Non-Darwinian Evolution. Science *164*, 788–798 (1969)

King, M. C., Wilson, A. C.: Evolution at two levels in human and chimpanzees. Science *188*, 107–116 (1975)

Sarich, V., Cronin, J. E.: Molecular systematics of the primates. In: Molecular anthropology. Genes and proteins in the evolutionary ascent of the primates. Goodman, M., Tashian, R. E., Tashian, J. (eds.), pp. 141–170. New York, London: Plenum Press 1976

Sokal, R. R., Michener, C. D.: A statistical method for evaluating systematic relationships. Univ. Kans. Sci. Bull. *38*, 1409–1438 (1958)

Tashian, R., Goodman, M., Ferrel, R., Tanis, R. J.: Evolution of carbonic anhydrase in primates and other mammals. In: Molecular anthropology. Genes and proteins in the evolutionary ascent of the primates. Goodman, M., Tashian, R. E., Tashian, J. (eds.), pp. 301–320. New York, London: Plenum Press 1976

Whitfield, H. J., Martin, R. G., Ames, B. N.: Classification of aminotransferase (C gene) mutants in the histidine operon. J. Mol. Biol. *21*, 335–355 (1966)

Wilson, A. C., Bush, G. L., Case, S. M., King, M. C.: Social structuring of mammalian populations and rate of chromosomal evolution. Proc. Nat. Acad. Sci., USA *72*, 5061–5065 (1975)

Wilson, A. C., Maxson, L. R., Sarich, V. M.: Two types of molecular evolution. Evidence from studies of interspecific hybridisation. Proc. Nat. Acad. Sci. USA *71*, 2843–2847 (1974a)

Wilson, A. C., Sarich, V. M., Maxson, L. R.: The importance of gene rearrangement in evolution: evidence from studies on rates of chromosomal, protein and anatomical evolution. Proc. Nat. Acad. Sci. USA *71*, 3028–3030 (1974b)

Wurster, D., Benirschke, K.: Indian muntjack, *Muntiacus muntjak:* A deer with a low diploid chromosome number. Science *168*, 1364–1366 (1970)

---

[7] The term "syntenic" is preferred to "linked", since this latter refers to genes that segregate together at meiosis with a significantly higher frequency than expected by Mendelian segregation. Not all "syntenic" genes may necessarily appear as "linked", if they are sufficiently distant in map units to allow recombination to occur frequently between them.

Chapter 12 Comparative Gene Mapping in Man and Other Primates

The Evolution of Chromosomes as Syntenic Groups

The Hominidae is a group showing remarkable evolutionary conservation at the chromosome level, in spite of the presumed chromosome rearrangements that have taken place during phylogeny. If chromosomes are conceived as syntenic groups rather than morphological entities, rearrangements such as inversions would not affect the number of gene loci per chromosome, but rather the relative position of genes within chromosomes. A telomeric fusion, such as the one postulated in the phylogeny of chromosome number 2 in man, would imply that two ancestral syntenic groups have become one in *Homo sapiens,* while remaining separate in the great ape species. It must be stated that none of the postulated rearrangements would necessarily cause genes to be transposed frome one non-homologous chromosome to another during phylogeny, as would be expected if translocation between non-homologous chromosomes had occurred. For this reason, the data supplied by chromosome banding and morphology lead us to expect a priori that syntenic groups in the Hominidae must have undergone only a limited amount of change.

In the past years, the development of powerful techniques for assigning single gene sequence to chromosomes has been greatly improved, and these have allowed studies of comparative gene mapping between phylogenetically related species to be undertaken. The most commonly used method consists of making heterokaryotic cell lines, by fusing cells of different species such as human (or ape) with rodent cells (e.g., mouse), using Sendai viruses or other agents. Such hybrid cells have been found preferably to lose human (or ape) chromosomes while retaining most of their rodent chromosome complement (Creagan and Ruddle, 1977). If a human (or ape) structural gene product were detected in such hybrid cells, it could be assigned to the retained human (or ape) chromosome(s) which can be identified by a chromosome-banding technique. Once a gene is assigned to a chromosome, any structural gene product which segregates in a cell line with the known marker is also assigned to that chromosome, or referred to as being "syntenic"[7]. Most of the available information on comparative gene mapping of single gene sequences has been obtained by somatic cell hybridisation, although other methods are available (Creagan and Ruddle, 1977); and some of them have also been used, such as adenovirus – 12 integration (Orkwiszewski et al., 1976). The results of these studies (see References) are summarized in Table 12.1, and they generally agree with the a priori expectation that syntenic groups have been conserved in the Hominidae. Some findings, however, deserve special comment. Firstly, there is good agreement between chromosome homology, as inferred by chromosome morphology and banding, and data on comparative gene mapping (see Table 12.1). Exceptions to this rule, however, do exist, one being that of chromosome 2 in man and its respective "arm homologues" in the great apes. Initial studies had shown that structural genes carried by HSA 2 were also carried by the two-arm homologue chromosomes in the chimpanzee *(P. troglodytes)*, but it could not be demonstrated whether genes carried by each arm of HSA 2 were the same as those carried by each corresponding "arm homologue" (Finaz et al., 1975). More recently, this problem has been solved showing that, contrary to expectations derived from presumptive (banding) homology, gene markers carried by HSA 2p, MDH-1, ACP-1 and GAL$^+$ ACT (Hamerton, Baltimore Conference, 1975), were assigned to PTR 13 (Sun et al., Winnipeg Conference, 1977), a chromosome recognized as homologue to HSA 2q. Pearson et al. (Winnipeg Conference, 1977) found similar results for MDH-1 in the chimpanzee, gorilla and orangutan homologue of HSA 2q (PTR 13, GGO 11, and PPY 11), whereas the gene locus for IDH, which had been located in HSA 2q, was assigned to PTR 12, GGO 12, and PPY 12, the presumed homologues to HSA 2p (Table 12.1). These findings indicated that "arm homologies" must be reconsidered in relation to chromosome 2 in man. Moreover, in the African green monkey, MDH-1 and IDH also segregated with different chromosomes (Pearson et al., 1977). It was also found that two loci which are in close proximity in the short arm of chromosome 2 in man (ACP-1 and MDH-1), showed dissociation in the chimpanzee and the orangutan, thus suggesting that their relative positions might not be strictly comparable to those found in HSA 2p. Another example of the lack of coincidence between "banding homologies" and "syntenic homologies" was the case of chromosome 20 in man and PTR 21, GGO 21, and PPY 21 in relation to ITP (Table 12.1; Estop et al., Winnipeg Conference, 1977).

Comparative gene mapping has succeeded in demonstrating interspecific chromosome homologies in situations in which the inference of homology by chromosome morphology and banding was not clearly

Table 12.1. Comparative gene mapping in the Hominidae and Cercopithecoidea

Man	Chimpanzee	Gorilla	Orangutan	Rhesus monkey	African green monkey
HSA 1	*PTR 1*	*GGO 1*	*PPY 1*	*MML A-1*	*CAE A-1* (CAE 4[b])
(p) ENO-1	+	+	+	+	+
(p) α-FUC	+				+
(p) AK-2	+			+	
(p) PGD	+ syntenic	+ syntenic	+ syntenic	+ syntenic	+
(p) PGM-1	+	+	+	+	+
					CAE A-7 (CAE 13[b])
(q) FH		+	+[a]	+	+
(q) GUK		+		+	
(q) PEP-C	+	+	+		+
HSA 2	*PTR 13*	*GGO 11*	*PPY 11*		
(p) MDH-1	+	+	+		
(p) ACP-1	+		+		
(p) GAL⁺ACT	+				
	PTR 12	*GGO 12*	*PPY 12*		
(q) IDH-1	+	+	+		
(q) GALT	+				
HSA 3	*PTR 2*	*GGO 2*	*PPY 2*	*MML A-2*	
GPX-1	+	+		+	
HSA 4	*PTR 3*	*GGO 3*	*PPY 3*	*MML A-6*	
PGM-2	+	+		+	
HSA 5	*PTR 4*	*GGO 4*	*PPY 4*	*MML A-4*	
HEX-B			+	+	
HSA 6	*PTR 5*	*GGO 5*	*PPY 5*	*MML A-3*	
SOD-2	+	+	+	+	
HSA 7	*PTR 6*	*GGO 6*	*PPY 10*		
β-GUS	+	+			
HSA 8	*PTR 7*	*GGO 7*	*PPY 6*	*MML B-1*	*CAE B-1*
GSR	+	+	+	+	+
HSA 9	*PTR 11*	*GGO 13*	*PPY 13*	*MML homologue?*	*CAE homologue?*
AL-1	+				AK-1 syntenic
ACO-S	+	+		ACO-S syntenic	ACO-S
AK-3	+	+		AK-3	
HSA 10	*PTR 8*	*GGO 8*	*PPY 7*		
GOT-1	+	+	+		
HSA 11	*PTR 9*	*GGO 9*	*PPY 8*	*MML C-1*	*CAE C-1*
LDH-A	+[c]	+	+	+	+
ACP-2	+				
HSA 12	*PTR 10*	*GGO 10*	*PPY 9*	*MML B-2*	*CAE B-2*
LDH-B	+	+	+	+	+
PEP-B	+	+	+	+	+
TPI	+	+	+	+	+
ENO-2	+				
GAPD	+	+	+	+	+
CS				+	+
HSA 13	*PTR 14*	*GGO 14*	*PPY 14*		
ESD		+			

Table 12.1. (continued)

Man	Chimpanzee	Gorilla	Orangutan	Rhesus monkey	African green monkey
HSA 14 NP	*PTR 15* +	*GGO 18* +	*PPY 15*		
HSA 15 MPI HEX-A PK$_{M2}$	*PTR 16* + + +	*GGO 15* + +	*PPY 16* + + +	*MML homologue?* HEX-A ⎫ syntenic PK$_{M2}$ ⎭	*CAE homologue?* HEX-A ⎫ syntenic PK$_{M2}$ ⎭
HSA 17 TK GALK AD-12	*PTR 19* + + +	*GGO 19* +	*PPY 19*		*CAE homologue?* TK ⎫ syntenic GALK ⎭
HSA 18 PEPA	*PTR 17* +	*GGO 16* +	*PPY 17*		
HSA 19 GPI	*PTR 20* +	*GGO 20* +	*PPY 20* +	*MML C-5* +	*CAE C-5* +
HSA 20 ITP	*PTR 21* (—) ITP and NP are syntenic	*GGO 21* (—) ITP and NP are syntenic	*PPY 21* (—)		
HSA 21 SOD-1	*PTR 22* +	*GGO 22* +	*PPY 22*		
HSA X HPRT G 6PD PGK α-GAL	*PTR X* + + +	*GGO X* + + + +	*PPY X* + + +	*MML X* + +	*CAE X* + + +

+ Means provisional or confirmed gene assignment.
(—) Means negative gene assignment.
(p) Short arm of chromosome.
(q) Long arm of chromosome.

[a] FH is assigned to PPY 1p, the homologous arm of HSA 1q (Garver et al. in Winnipeg Conference, 1977).

[b] CAE 4 and CAE 13 refer to the classification of Finaz et al. (1977a, b) whereas CAE A-1 and CAE A-7 refer to the classification of Stock and Hsu (1973). This latter classification is used for the Rhesus monkey (MML) and the African green monkey.

[c] LDH-A is assigned to PTR 9q (de Grouchy et al. in Baltimore Conference, 1975).

This table has been composed from data presented in the Baltimore Conference (1975), the Winnipeg Conference (1977), and the following references: Finaz et al. (1975, 1977a, b); Rebourcet et al. (1975); Orkwiszewski et al. (1976); Cochet et al. (1977); Garver et al. (1977); de Grouchy et al. (1977); and van Cong et al. (1978).

evident. This was the case of chromosome 9 in man that had no recognizable homologue in the gorilla and the orangutan (Paris Conference, 1971 supplement 1975). Pearson et al. (1977) reported that two gene loci (ACO-S and AK-3), which segregate with chromosome 9 in man and with its corresponding homologue in the chimpanzee (PTR 11), also segregated with an acrocentric chromosome in the gorilla (GGO 13), a chromosome that was not recognized as being homologous to any chromosome in man or other ape species (Paris Conference 1971 supplement 1975). Moreover, similar syntenic groups to HSA 9 were also found in the orangutan, the Rhesus monkey and the African green monkey, although no chromosome assignments were made. In these species, as in the gorilla, there is not a single chromosome pair with a morphology and banding pattern to resemble HSA 9, thus indicating that chromosome 9 in man must have been of more recent origin than those other chromosomes showing a remarkable evolutionary conservation, such as the X chromosome.

Finally, chromosome 14 in man had no presumptive homologue in the gorilla (Paris Conference, 1971 supplement 1975), but Pearson et al. (1977) proposed that GGO 18 should be considered as its corresponding homologue based on the fact that this chromosome segregated with NP. However, it should also be mentioned that the findings reported by Pearson et al. (1977) in relation to the syntenic homology between HSA 9 and GGO 13, and HSA 14 and GGO 18, contradict previous reports of the same research group, who had previously postulated different chromosome homologies between man and the gorilla (see Meera Khan et al., 1975). In this report, ITP had been assigned to chromosome 13 in the gorilla; and NP, ITP, and chromosome 13 had been previously found to segregate together in seven hybrids, and in 9 of 10 subclones derived from one of these clones, positive for such markers. This and the fact that GGO 13 somewhat resembled HSA 14 in morphology and banding pattern, led Pearson et al. (1975) to propose that GGO 13 was the corresponding homologue to HSA 14 (see Baltimore Conference, 1975).

The Conservation of the Syntenic Groups Among the Hominidae and Cercopithecoidea

Chromosome homologies, as inferred by banding, have also been observed between the Hominidae and some species of Cercopithecoid monkeys such as the Rhesus monkey *(Macaca mulatta)*, the African green monkey *(Cercopithecus aethiops)* and species of baboons *(Papio papio, Papio cynocephalus)*; (Stock and Hsu, 1973; Finaz et al., 1977a, b; Garver et al., 1977; de Grouchy et al., 1977). Paradoxically, these Cercopithecoid species show greater chromosome homology to the Hominidae than do the Hylobatidae, and this similarity is good evidence of a remarkable evolutionary conservation at the chromosome level in the catarrhine lineage. Comparative gene mapping between the Hominidae and the above-mentioned species of Cercopithecoid monkeys (Garver et al., 1977; Pearson et al., 1977; Winnipeg Conference, 1977; see Table 12.1) confirmed that some chromosomes had remained remarkably stable during phylogeny, a period which roughly encompassed 35 M.y. since the Hominoid-Cercopithecoid divergence. For example, SOD-2, which had been previously assigned to chromosome 6 in man, not only was assigned to PTR 5, GGO 5, PPY 5 but also to its corresponding homologue in the Rhesus monkey. LDH-A was found in the corresponding homologue to HSA 11 in the three species of great apes and two Cercopithecoids (Rhesus and African green monkey), and the syntenic group LDH-B, PEP-B, TPI, and GAPD in their corresponding homologue to HSA 12. Other genes such as CS were also found in this syntenic group, although not in all the species tested (see Table 12.1). The X chromosome markers G-6PD, α-Gal and PGK were also syntenic in all species except for PGK in PPY X and the Rhesus monkey X chromosome where electrophoretic separation between primate and Chinese hamster forms of PGK was not achieved (Garver et al., 1977).

Comparative Gene Mapping Between Hominidae-Cercopithecoidea and the Possible Origin of Chromosome 1 in Man

The presumptive chromosome rearrangements that could have occurred in the phylogeny of chromosome 1 in man deserve special comment, and they can be inferred from data on comparative gene mapping between the Hominidae and the Cercopithecoidea. In the first place, chromosome 1 in man has a metacentric homologue in the great apes, the Rhesus monkey and the baboons (*P. papio* and *P. cynocephalus*), in which the largest chromosome in their respective complements consists of a metacentric chromosome whose arms show similar G-band patterns to the human chromosome 1. In man, chromosome 1 has a secondary constriction which is absent in all other primate species; the appearance of such a constriction at the proximal region of one arm has increased the length of this arm, thus making it the longer of the two. In the other primate species in which the secondary constriction is not present, the arm ratio is therefore the reverse of that in man. Thus in the great apes, the Rhesus monkey and the baboons the *short* arm of their chromosome 1 is homologous to HSA 1q, and the *long* arm of their chromosome 1 is homologous to HSA 1p. G-band or R-band homologies between chromosome arms are very evident between man and the great apes (Paris Conference, 1971 supplement 1975), but G- or R-banding patterns between HSA 1q and the baboons and the Rhesus monkey 1p can be better matched if we assume that a paracentric inversion had occurred between them at some stage during phylogeny (Finaz et al., 1977a, b; de Grouchy et al., 1977). In the African green monkey (CAE), on the other hand, chromosome 1 in man has no direct metacentric homologue; but it can be derived from a presumptive fusion of a large subtelocentric chromosome, CAE 4 (homologous to 1p) with a smaller acrocentric chromosome, CAE 13 (homologous to 1q). Here again, G- and R-band homologies between HSA 1p and CAE 4 match well, but the banding pattern of HSA 1q and CAE 13 are better matched if we assume that a paracentric inversion has occurred between them at some stage during phylogeny. The presumed paracentric inversion between CAE 13 and HSA 1q is the same as the one between HSA 1q and the baboon or Rhesus 1p, so that chromosome 1 in the baboon or the Rhesus can be directly derived by a simple fusion of CAE 4 and CAE 13 (Stock and Hsu, 1973; Finaz et al., 1977a, b; de Grouchy et al., 1977; see Fig. 12.1).

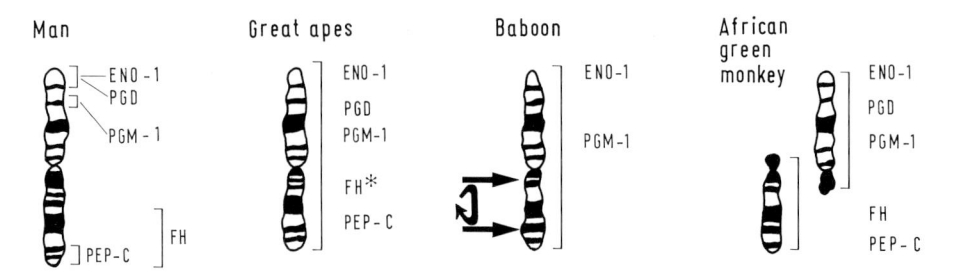

Fig. 12.1. Comparative gene mapping of chromosome 1 in man and its homologues in the great apes (excluding the pygmy chimpanzee), the baboon, and the African green monkey, according to Finaz et al. [Annales de Génétique 20, 85–92 (1977)], and Garver et al. [Chromosomes Today 6, 191–199 (1977)]. In this latter report data is presented on chromosome 1 in the Rhesus monkey instead of baboon (see Table 12.1). Ape and baboon chromosome 1 are inverted. *Straight arrows* pointing to the baboon chromosome 1 show presumptive chromosome breaks, and *curved arrow* indicates the paracentric inversion from which ape and human chromosome 1 can be derived. *Small arrows* pointing to chromosome 1 in man show the sites where genes are contained as indicated by *vertical bars*. FH* has been assigned to chromosome 1 of the gorilla and the orangutan only (see Table 12.1)

Now let us consider the gene loci contained in each of these chromosomes as reported by Garver et al., 1977; Finaz et al., 1977 a, b; and de Grouchy et al., 1977; and comment on the possible implications on the phylogeny of chromosome 1 in man (Fig. 12.1). In man, HSA 1p contains PGD, PGM-1 and ENO-1, whereas FH and PEP-C are in HSA 1q. In the three species of great apes studied (*Pan paniscus* excluded) ENO-1, PGM-1 and PEP-C were assigned to chromosome 1, and FH to GGO 1 and PPY 1. No assignments were made to chromosome arms except for FH in PPY 1. In the baboon *(Papio papio)* No. 1, ENO-1, and PGM-1 were also found to be syntenic, but no arm assignment was made. It must be remarked, however, that in *Papio cynocephalus* Warburton et al. (1975) had localized the 5S rDNA sequences at the terminal region of the short arm of chromosome 1, thus coinciding with the human and great ape location of these sequences. This finding suggested that the terminal region of this chromosome in the baboon must have been conserved, a situation which is comparable with the break points of

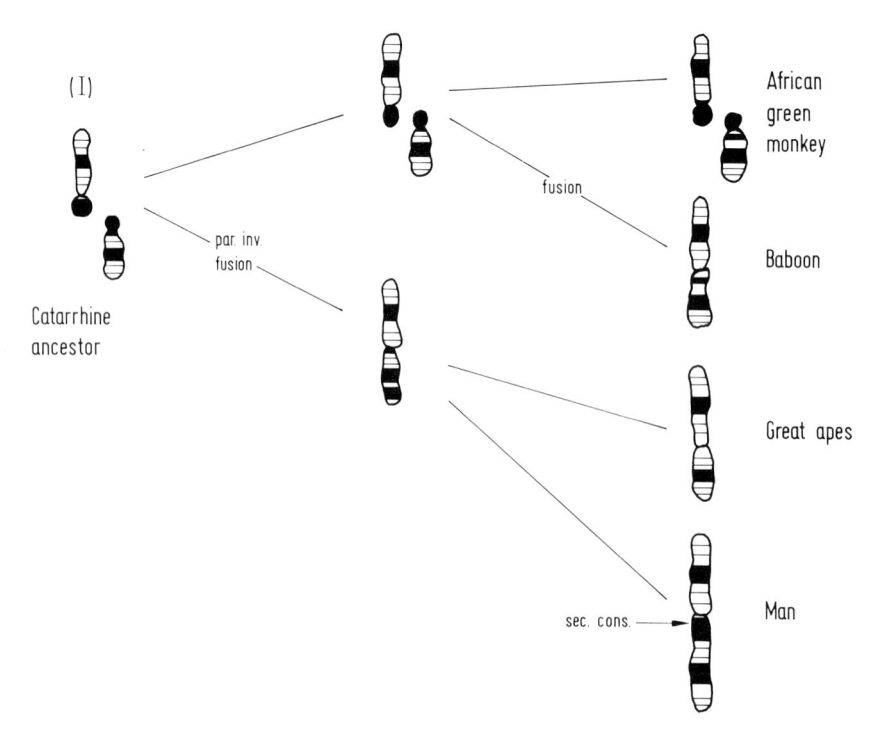

Fig. 12.2. 1st Hypothesis (I). Derivation of chromosome 1 from a Cercopithecoid-Hominoid (Catarrhine) ancestor with chromosome 1 similar to the African green monkey following Finaz et al., Annales de Génétique 20, 85–92 (1977). (*sec. cons.*, secondary constriction; *par. inv.*, paracentric inversion)

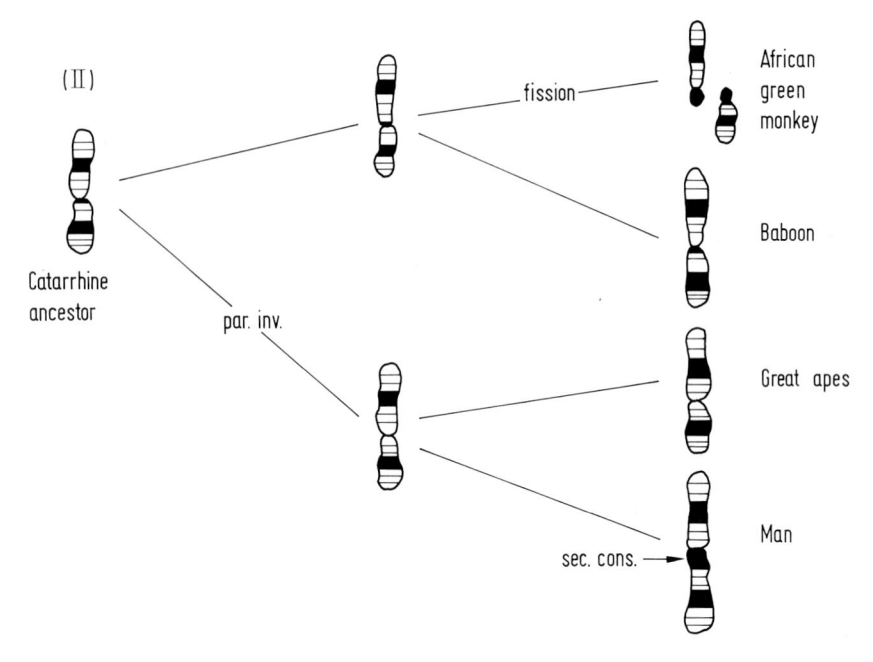

Fig. 12.3. 2nd Hypothesis (II). Derivation of chromosome 1 from a Cercopithecoid-Hominoid (Catarrhine) ancestor with chromosome 1 similar to the baboon following Finaz et al., Annales de Génétique 20, 85–92 (1977). (*sec. cons.*, secondary constriction; *par. inv.*, paracentric inversion)

the paracentric inversion postulated by Finaz et al., 1977 a, b; and de Grouchy et al., 1977. Finally, in the African green monkey, CAE 4 was found to contain PGD, ENO-1 and PGM-1 (HSA 1p markers), whereas CAE 13 contained PEP-C and FH (HSA 1q markers). Findings of this kind led to the proposition of three different hypotheses in relation to the origin of chromosome 1 in man. One is that the African green monkey represents the ancestral condition in the catarrhini (Fig. 12.2), and that one paracentric inversion took place in the line to the Hominoidea together with a fusion, while another fusion occurred in the baboon lineage. The second (Fig. 12.3) postulated that the ancestral chromosome was similar to PPP 1 and that one fission took place in the lineage leading to the green monkey, and one paracentric inversion took place in the line to the Hominoidea. The third hypothesis (Fig. 12.4) was that the ancestral chromosome was similar to that of the great apes, and that chromosome rearrangements exclusively took place in the Cercopithecoid lineage (a paracentric inversion), and within the Cercopithecoid radiation (a fission).

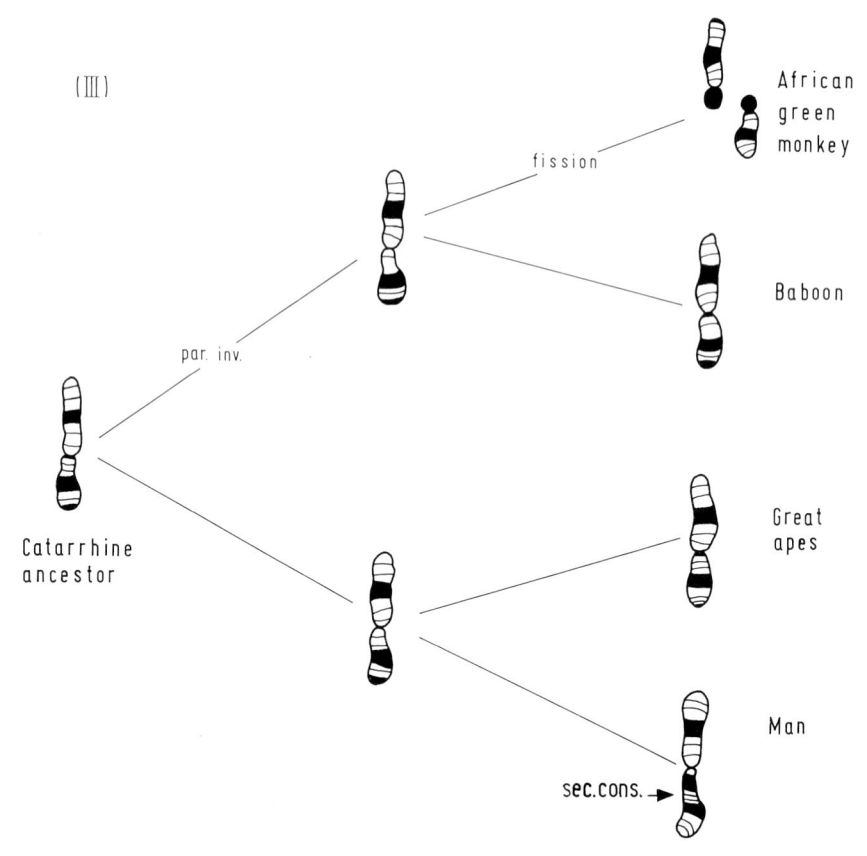

Fig. 12.4. 3rd Hypothesis (III). Derivation of chromosome 1 from a Cercopithecoid-Hominoid (Catarrhine) ancestor with chromosome 1 similar to the great apes following Finaz et al., Annales de Génétique 20, 85–92 (1977). (*sec. cons.*, secondary constriction; *par. inv.*, paracentric inversion)

Are Chromosomes Frozen Accidents?

The findings reported in the previous section supply good evidence that chromosomes may be conserved during relatively long periods of evolutionary divergence. One of the possibilities for such a remarkable evolutionary conservation may be the fact that chromosomes have attained a certain degree of functional capacity resulting from a qualitative/quantitative DNA content, as well as internal organization, which does not tolerate anything but limited change. Ohno (1973) has commented on the fact that chromosomes could thus have become "frozen accidents" as has been well demonstrated in the case of the X chromosome in mammals, in which there is striking similarity between species. It has been shown (Ohno, 1976) that the X chromosome of many mammalian species, some of them phylogenetically distant, such as man, chimpanzee, Rhesus monkey, rabbit, dog, pig, red deer, and mouse, represent about 5% of the total genome. They show a remarkably similar submetacentric morphology and banding pattern. However there are some exceptions such as the mouse and the red deer, in which the X is acrocentric. In some other species, e.g. the Syrian hamster and the reindeer, the size of the X chromosome has increased above the average DNA content of most mammals. Nonetheless such increase has been shown to be due to constitutive heterochromatin, or genetically inactive DNA. Perhaps the best proof of the evolutionary conservation is supplied by comparative gene mapping which has shown that human X linked markers are also found in the X chromosomes of other species, e.g., G-6PD in man, the great apes, the Rhesus monkey, the African green monkey, the horse, the donkey, hare, mouse, and kangaroo. Results of this kind indicate that the X chromosome in mammals has remained basically unmodified for at least 125×10^6 years, a period which encompassed the phyletic divergence of mammals.

In relation to autosomes, the existence of frozen accidents is more difficult to demonstrate, since there are few species in which gene mapping has been undertaken. The available data, apart from man and the primates mentioned, accounts for studies in the mouse, the Norway rat, the rabbit, the deer mouse, the guinea pig and the Indian muntjack, but there is a considerable dearth of information over a wide range of species. Searle (1975) has suggested that during the mammalian radiation in which extensive genome reshuffling took place by chromosome rearrangement, some chromosome regions might, however, have remained intact either due to functional necessity or chance. For example, the albino (c) and the pink eye dilute (p) loci are equally distant in mouse, rat, and deer mouse, and they are probably linked in the Syrian hamster, while rat and mouse have the haemoglobin β-chain locus (Hbb) closely linked to (c), and the two species have the locus for resistance to warfarin in the same position relative to (c) and to Hbb. On the other hand, the linkage group involving the gene for brown pigmentation (b), showed little indication of having been conserved in the rodents, although both mouse and rat showed linked loci for anaemia (Pgm-2 and Pgd). It is interesting to remark that in man, the homologous loci (PGM-1 and PGD) are also syntenic, and have been assigned to the short arm of chromosome 1. Furthermore, a comparison between human and mouse linkage groups has shown that in these two species the loci for pancreatic and salivary α-amylase were closely linked, and the four heavy chain loci (Ig-1 to Ig-4), which in the mouse were closely linked, were closely linked in man and in the rabbit as well. However, there is ample evidence of syntenic groups in mouse which are not syntenic in man (Searle, 1975); an example is that of Id-1 and Dip-1 in the mouse which are contained in chromosome 1, whereas the homologous loci in man, ICD-S and PEP-C, respectively, are contained in chromosome 2 and 1 respectively. The important conclusion from such comparisons is that the conservation of syntenic groups in mammals is more obvious between phylogenetically closely related species, e.g., rodents, than between rodents and man. However, for very closely linked loci there is more evidence in favour of their evolutionary conservation, even in phylogenetically distant species, than for their separation, probably due to the fact that their distance had been so very close as to make separation by chromosome rearrangement very rare, or that if separation ever occurred it had led to less fit organisms being eliminated by selection.

Gene Duplication, Polyploidy, and Evolutionary Frozen Chromosomes

The fact (a) that eukaryotes have extensively increased their genome size in relation to prokaryotes, (b) that auto-tetraploid and auto-octaploid frog species had been reported in South America (Saez and Brum, 1960; Beçak et al., 1966, 1967), and (c) the observation that many structural genes which have probably arisen by gene duplication (e.g., α- and β-globin, LDH-A and LDH-B, γ heavy and γ light chains) have not been found to be linked in man, led Ohno (1970) to suggest that polyploidy could have occurred at some early stage of vertebrate radiation. Moreover, the findings in the fish of the suborder Salmonidae indicated that tetrapolyploidy, having resulted from a duplication of a diploid organism, had been subsequently followed by chromosome rearrangement (e.g., Robertsonian fusions) which accounted for differences between chromosome number between, and even within, individuals of the same species (Ohno et al., 1965). It must be remembered, however, that polyploidy could only have occurred as an effective mechanism in increasing genome size, at a stage of vertebrate evolution when the chromosomal sex-determining mechanisms had not yet been developed (Ohno, 1970). In a XX-/XY sex system, for example, a duplication of the entire chromosome set would result in XXXX females and XXYY males, the latter producing XY-bearing spermatozoa as a result of disjunction of one X and one

Y from their respective homologues in meiosis. Offspring of such crosses would all be XXXY and, if the Y were capable of determining a male phenotype, no female offspring would be produced. If not, only sterile intersex offspring would result.

Since the development of chromosomal sex-determining mechanisms appeared at the time of the emergence of the reptiles from the common vertebrate stock, or approximately 280×10^6 years ago, polyploidy could have occurred in an early vertebrate ancestor, no later than the fish–amphibian stage. If this assumption were correct, the chromosome complement of reptiles, birds, and mammals could then be derived from an ancestral diploid chromosome set of a primitive vertebrate which became duplicated (autotetraploidy), and which later underwent subsequent "diploidization" by chromosome rearrangement. If we imagined a tetraploid organism with sets of four homologous chromosomes (quadruplets) instead of pairs, "diploidization" as envisaged by Ohno (1970) could occur if two chromosomes in each quadruplet were capable of acquiring an identical kind of rearrangement (e.g., an inversion) so that bivalents (and not quadrivalents) would be formed at meiosis. Thus, each quadruplet will be divided into two pairs of homologous chromosomes. However, since the two pairs of homologous chromosomes derived from each quadruplet would have had a common origin, we would expect that, in terms of morphology and synteny, they would resemble each other more closely than chromosome pairs derived from other quadruplets. Moreover, if the amount of chromosome change involved in "diploidization" had been limited we would be able to infer which chromosome pairs had originated from an ancestral "quadruplet". Comings (1972) analysed the human chromosome complement with banding techniques and grouped pairs of chromosomes into presumptive arrangements such as: (1+2), (3q+3p), (4+5), (6+13 inverted), (7+8), (9+10), (11+12), (14+15), (16+17), (19+20), and (21+22); chromosome pair 18, the X and the Y chromosome not being assigned to any group. We must point out, however, that such comparison may be biased by the fact that it was done on chromosomes of species now existing, in which there has been considerable chromosome change during phylogeny. In the first place the amount of change which has occurred between chromosome pairs of a quadruplet *within* species has not been equal. Let us consider human chromosome 11 and 12 which show similar morphology as demonstrated by banding, and have been found to carry genes for very similar products (LDH-A in 11 and LDH-B in 12), a fact that points to their possible common origin. Let us now consider the corresponding

thus resembling human satellite I DNA. This satellite, however, has not been characterized.

Finally, it is important to mention that the proportion of DNA showing a low frequency of repetition (from 1000 to 10 copies), which can be kinetically defined as ranging from Co.t 10^0 to 10^2, was also found to vary among the Hominoidea, Cercopithecoidea and Lorisoidea (Gumerson, 1972). The Hominoidea, for example, had roughly one third of their repetitive DNA within this Co.t range, while the Cercopithecoidea had most of their repetitive DNA at values below 10^0. In the Lorisoidea, in which similar sequence families were found, a small percentage of their repetitive DNA reassociated between a Co.t range of 10^0 to 10^1.

Repetitive DNA in Man. Evidence of its Evolutionary Conservation

Although repetitive DNA is mainly composed of apparently functionless sequences that may thus be allowed to evolve at considerably faster rates than single copy-DNA, there is good evidence of its evolutionary conservation. Jones (1977) has discussed how a fraction of Rhesus monkey repetitive DNA may form thermally stable heteroduplexes with baboon and human DNA respectively. However, the fraction involved in heteroduplex formation was found to decrease with increasing evolutionary distance of the species tested from the Rhesus monkey. Deininger and Schmid (1976) have analysed the thermal stability of human and chimpanzee (spp. *troglodytes*) DNA heteroduplexes. These heteroduplexes were formed (1) between highly and intermediate repetitive human DNA with highly and intermediate repetitive chimpanzee DNA, and (2) between single-copy human DNA with single copy chimpanzee DNA. Using human homoduplexes as criteria of relatedness, it was shown that the single-copy heteroduplex DNA had a lower thermal stability than the corresponding human single-copy homoduplex (see also Chap. 10), but the repetitive heteroduplex had a melting profile undistinguishable from the repetitive human homoduplex. This result indicated that the highly and intermediate repetitive DNA components of man and the chimpanzee were almost identical, showing a greater degree of similarity than the single-copy DNA. This could have resulted from two alternatives: one, that the repetitive sequences present in both species were formed prior to their phyletic divergence and conserved ever since within limited change; the other, that families of repetitive DNAs in these two species consisted of related base sequences having changed a similar proportion of their bases at random sites.

Jones and Purdom (1975) and Jones (1977) analysed the thermal stability of DNA heteroduplexes formed between human satellite cRNA III and genomal DNA of man, chimpanzee and orangutan. The T_m value of these heteroduplexes was, respectively, 70 °, 63 °, and 59 °C, a finding that indicated increasing sequence divergence with increasing evolutionary distance from man. Moreover, homologous sequences to human satellite III were by this method undetectable in the genome of other primates phylogenetically further away from man than the orangutan, such as the gibbon or the baboon. These results indicated that the base sequence of human satellite III DNA must have been present in the human lineage as far back as the emergence of the Hominidae, subsequent to the branching off of the Hylobatidae. However, they did not indicate whether this sequence was already amplified in the common ancestor we share with the great apes, or happened after the splitting of the Hominidae into different species.

Finally, the level of hybridisation of the complementary RNA to chimpanzee satellite A (cRNA A) was found to be higher with those fractions of human or chimpanzee DNA with buoyant densities of 1.969 g/ml than with any other fraction in a gradient (Prosser et al., 1973). Similar results were obtained by hybridising human satellite cRNA III to density gradients of human or chimpanzee DNA. These results together with the previous findings of similar physical characteristics between chimpanzee satellite A DNA and human satellite III DNA, and the finding that they showed a similar chromosome distribution in both species were good evidence in favour of sequence homology between these satellites. Interspecific in situ hybridisations between each human satellite cRNA I, II, III and IV with chromosome preparations of the great ape species (Chap. 14), supply further evidence of the evolutionary conservation of these four human satellite DNAs.

Satellite DNAs in Man and Other Organisms. Possible Explanations of Their Evolutionary Conservation

Although good evidence exists for the evolutionary conservation of the four major satellite DNAs in man, the reason why satellite DNA might be conserved is still unclear. In the first place, little is known about the internal structure of satellite DNAs in man as revealed by restriction endonuclease digestion, and nothing is known about their base sequence. Restriction digests of human satellite III DNA with Hae III produced fragments ranging from 200 base pairs in multiples of this number up to approximately 2200 base pairs, showing differences in agarose gel band patterns between male and female DNA. This was due to a male specific repetitive DNA fraction isolated with satellite III which was approximately 3500 base pairs long, and shown to be confined to the human Y chromosome in the non-fluorescent region of the long arm and in the proximal part of the brilliant fluorescent portion (Bostock et al., 1978). A restriction pattern of human satellite DNA and satellite and repetitive DNA of other mammalian species such as mouse, sheep, calf, and rat, showed repeat periodicities with one another and with the restriction pattern observed with the α-satellite of the African green monkey (Maio et al., 1977), where the digest was composed of monomer units of approximately 176 ± 4 base pairs. In this latter species, the restriction sites of the α-satellite fell between adjacent nucleolosomes, producing similar-sized monomers with different endonucleases (Musich et al., 1977), a finding that suggested that periodicities could have resulted from unequal crossovers in nucleolosome interstices.

It must be mentioned that although there is no information on the base sequence of satellite DNAs in man, two of the four human satellite DNAs (II and III), were found to cross-hybridise in vitro (Melli et al., 1975). Since one of them (satellite II) is always isolated from the heavy side of the main band DNA in a $Ag^+Cs_2SO_4$ gradient, and the other (satellite III), on the light side, the possibility of hybridisation resulting from contaminated fractions was highly unlikely. Thus, cross-hybridisation may well indicate that one sequence could have derived from the other by a mechanism of amplification and divergence of a previously existing repetitive sequence, as proposed by Southern (1970). Satellite II has a ΔT_m of 2 °C compared to 10 °C for satellite III (Chap. 9), so that the former shows a lower degree of internal heterogeneity than the latter. This finding together with (1), lack of detectable hybridisation between satellite cRNA II and chromosomes of *Pan troglodytes* against detectable hybridisation of cRNA III with chromosomes of *Pan, Gorilla* and *Pongo;* and (2), the finding of a satellite DNA fraction similar to human satellite III in *Pan troglodytes,* but not to human satellite II, led Jones et al. (1972) and Jones (1976, 1977) to postulate that satellite II had derived from satellite III at a later stage during phylogeny in the line leading to man, after the divergence of the ancestors of man and the chimpanzee. This hypothesis assumed, however, that the rate of nucleotide substitution in each satellite DNA sequence was constant during phylogeny, so that internal heterogeneity of satellite sequences would be indicative of their relative age in relation to one another. This might not hold true if the rates of nucleotide substitution differed among different satellite DNAs. It also assumed that satellite II must be a repetitive sequence unique to the human genome, and therefore absent from the genome of *Gorilla* and *Pongo,* as it was found to be from *Pan* (Jones et al., 1972). Interspecific in situ hybridisations have shown (Chap. 14) that human satellite cRNA II hybridised with chromosomes of *Gorilla* and, moreover, with *Pongo,* under stringent conditions of RNA–DNA hybridisation that would make cross-reaction between cRNA II and homologous sequences to human satellite III DNA very unlikely to occur. Moreover, the distributions of hybridisation of human cRNA I, II, III and IV in the chromosome complement of man and the great apes suggested that the derivation of one satellite sequence from another previously amplified sequence was unlikely, a point that will be fully dealt with in Chap. 14.

Finally, it is important to discuss why apparently functionless DNA might have been conserved during phylogeny, and whether such remarkable conservation could have been due to a possible role of satellite DNA in evolution. In other organisms, such as in reptiles and birds, for example, there is good evidence that satellite DNA may have played a role in the development of sex chromosome dimorphism (Singh et al., 1976). In these two groups, a dimorphic sex chromosome pair can be found in the female (ZW), against the homomorphic male sex (ZZ). However, sex chromosome dimorphism is absent from the most primitive snakes (the Boidae), whereas in other more evolved snakes (e.g., *Ptyas mucosus* family Colubridae) the sex chromosomes are morphologically identical in size and shape. They can be distinguished from one another only by their intensity of staining with C-banding or by comparing their pattern of late DNA replication. In the most advanced snakes (e.g., *Elephe radiata*), the sex chromosomes are clearly dimorphic. In this species, female and male DNA differed owing to a satellite DNA fraction which was localized in the W chromosome of female animals by in situ hybridisation. The complementary RNA (cRNA) to this W-specific satellite not only hybridised with the W chromosome of other snake species, but also with the W chromosome of the domestic chicken *(Gallus domesticus),* a finding which indicated a remarkable evolutionary conservation of both the W-satellite DNA with the W chromosome. In situ hybridisation with the chromosomes of two species of Boidae showed, on the other side, no W-specific satellite detectable in any chromosome. However, W-specific satellite cRNA hybridised to one of the two homomorphic sex chromosomes in *Ptyas mucosus*. These experiments demonstrated that the appearance of a W-specific satellite DNA anteceded the morphological differentiation of the W from the Z chromosome, so that sex chromosome dimorphism must have been a consequence rather than a cause of sex chromosome differentiation. A possible mechanism by which such morphological differentiation could have arisen was, according to Singh et al. (1976), the suppression of crossing-over between sex chromosomes in meiosis due to the amplification in one of them (the W) of a repetitive DNA sequence absent in its homologue (the Z chromosome). Thus, the previously homologous sex chromosomes would then have been allowed to evolve separately and eventually become morphologically distinct.

Among mammals, the species of the genus *Dipodymis* (kangaroo rats) also show evidence of a possible evolutionary role of satellite DNA. By comparing chromosome morphology with satellite DNA content in chromosomes, Hatch et al. (1976) suggested that satellite DNA had been probably involved in chromosome rearrangement within the group, which in turn might have resulted in speciation. In *D. ordii*, one of the three satellite DNAs (HS-α) was found to occur in at least 12 sequence variants in different amounts in the genome that could be derived from a 6 base sequence 5' TTAGGG 3' by one or two nucleotide substitutions (Fry and Salser, 1977). However, while none of these sequences was related to the other two satellite DNAs present in the genome of the same species, and in other species of the same genus (Salser et al., 1976), the base sequence of the HS-α-satellite was identical with that of the α-satellite of the guinea pig (Southern, 1970). Moreover, this sequence was similar to that found in the α-satellite of the pocket gopher, *Thomomys bottae,* and the antelope ground squirrel, *Ammospermophilus leucurus*. Since the species compared belonged to different rodent sub-orders which had diverged from one another approximately 40–50 million years ago, the evolutionary conservation of this sequence could have resulted from one of two alternatives. One was chance convergence, which is highly unlikely; the other, proposed by Fry and Salser (1977), was that each species possessed a set, or "library", of sequences, whose members might have been independently amplified in each species, and maintained ever since by the acquisition of unspecified functions which must have conferred a selective advantage.

A complete explanation of why satellite DNAs have been conserved will await further investigation before the function of repetitive DNAs, both during ontogeny and phylogeny, is fully understood. However, a better understanding of how satellite DNAs in man and homologous sequences in the great apes have evolved can be obtained by studying their chromosome distribution. Since these sequences could have been amplified either before or after speciation, a comparison of the relative amounts and chromosome distribution of satellite DNAs common to different species is illuminating to distinguish between these two possibilities. This aspect will be fully dealt with in the next chapter.

References

Bostock, C., Gosden, J.R., Mitchell, A.R.: Localisation of a male specific DNA fragment to a sub-region of the human Y chromosome. Nature (London) *272*, 324–328 (1978)

Deininger, P.L., Schmid, C.W.: Thermal stability of human DNA and chimpanzee DNA heteroduplexes. Science *194*, 846–848 (1976)

Fry, K., Salser, W.: Nucleotide sequences of HS-α-satellite DNA from kangaroo rat *Dipodomys ordii* and characterization of similar sequences in other rodents. Cell *12*, 1069–1084 (1977)

Gummerson, K.S.: The evolution of repeated DNA in primates. Thesis, John Hopkins Univ. 1972

Hatch, F.T., Bodmer, A.J., Mazrimas, J.A., Moore, D.H.: Satellite DNA and cytogenetic evolution DNA quantity, satellite DNA and karyotypic variations in kangaroo rats (genus *Dipodomys*). Chromosoma (Berlin) *58*, 155–168 (1976)

Jones, K.W.: Repetitive DNA sequences in animals. Proceedings of the Leiden Chromosome Conference (1974) In: Chromosomes today. Pearson, P.L., Lewis, K.R. (eds.), Vol. 5, pp. 305–313. New York: John Wiley & Sons. Jerusalem: Israel Univ. Press 1976

Jones, K.W.: Repetitive DNA and primate evolution. In: Molecular structure of human chromosomes. Yunis, J.J. (ed.), pp. 295–326. New York: Academic Press 1977

Jones, K.W., Prosser, J., Corneo, G.; Ginelli, E., Bobrow, M.: Satellite DNA constitutive heterochromatin and human evolution. In: Modern aspects of cytogenetics: constitutive heterochromatin in man. Symposia Medica Hoechst. Pfeiffer, R.A. (ed.), Vol. 6, pp. 45–61. Stuttgart, New York: Schattauer-Verlag 1972

Jones, K.W., Purdom, I.F.: Evolution of defined classes of human and primate DNA. In: Chromosome variations in human evolution. Boyce, A.J. (ed.), pp. 39–51. London: Taylor and Francis 1975

Kurnit, D.M., Maio, J.J.: Variable satellite DNA's in the African green monkey *Cercopithecus aethiops*. Chromosoma (Berlin) *45*, 387–400 (1974)

Maio, J.J.: DNA strand reassociation and polynucleotide binding in the African green monkey. J. Mol. Biol. *56*, 579–595 (1971)

Maio, J.P., Brown, F.L., Musich, P.R.: Subunit structure of chromatin and the organization of eukaryotic highly repetitive DNA. Recurrent periodicities and models for the evolutionary origins of repetitive DNA. J. Mol. Biol. *117*, 637–655 (1977)

Melli, M., Ginelli, E., Corneo, G., di Lernia, R.: Clustering of the DNA sequences complementary to repetitive nuclear DNA of HeLa cells. J. Mol. Biol. *93*, 23–38 (1975)

Musich, P.R., Maio, J.J., Brown, F.L.: Subunit structure of chromatin and the organization of eukaryotic highly repetitive DNA: indications of a phase relation between restriction sites and chromatin subunits in African green monkey and calf nuclei. J. Mol. Biol. *117*, 657–677 (1977)

Prosser, J.: Satellite DNA in man and three other primates. Thesis, Univ. Edinburgh (1974)
Prosser, J., Moar, M., Bobrow, M., Jones, K.W.: Satellite sequences in chimpanzee *(Pan troglodytes)*. Biochim. Biophys. Acta *319*, 122–134 (1973)
Salser, W., Bowen, S., Browne, D., El Adli, F., Fedoroff, N., Fry, K., Heindell, H., Paddock, G., Poon, R., Wallace, B., Whitcome, P.: Investigation of the organization of mammalian chromosomes at the DNA sequence level. Fed. Proc. *35*, 23–35 (1976)
Singh, L., Purdom, I.F., Jones, K.W.: Satellite DNA and evolution of sex chromosomes. Chromosoma (Berlin) *59*, 43–62 (1976)
Southern, E.M.: Base sequence and evolution of guinea-pig α-satellite DNA. Nature (London) *227*, 794–798 (1970)

Chapter 14   The Chromosome Distribution of Homologous Sequences to the Four Human Satellite DNAs in the Hominidae

The Distribution of Satellite I, II, III and IV in the Human Chromosome Complement

Techniques of in situ hybridisation, developed by Gall and Pardue (1969), have enabled us to detect and localize repetitive DNA sequences in chromosomes. This method has frequently been applied to study the chromosome distribution of repetitive DNA sequences in man. Each human satellite DNA has been used as a template for in vitro synthesis of a radioactively complementary RNA (cRNA), using a bacterial DNA-dependent RNA polymerase. Each cRNA was then applied to denatured human chromosome preparations, and the sites where stable cRNA-DNA hybrids were formed were subsequently identified with autoradiography. Different amounts of these hybrid molecules, both in vitro and in situ, may be formed according to the temperature of incubation (Moar et al., 1975), because each of the three human satellite DNAs (I, II, and III) proved to have an optimum temperature (T_{OPT}) at which cRNA-DNA hybridisation was most effective. Above their respective T_{OPT}, the amount of hybridisation detected in vitro and in situ decreased.

In 1971 Jones and Corneo initially reported the distribution of satellite II in the human chromosome complement, while that of satellite III and I was later reported by Jones et al. in 1973 and 1974, respectively. These studies showed that satellite II was localized in the secondary constriction region of chromosome 1 and 16 in man, and to a lesser extent also in chromosome 9, while other heterochromatic regions, such as those of the centromere of the acrocentric chromosomes, were only minor sites of hybridisation. High concentrations of satellite III were detected in the secondary constriction of chromosome 9, and smaller amounts of this sequence were found in the centromere regions of the acrocentric chromosomes, in chromosome 16, and occasionally in the centromere region of chromosome 1. Satellite I was mainly found in the Y chromosome long arm, while only minor sites of hybridisation were detected in some centromere regions of the human chromosome complement. A greater definition of this technique was achieved by using chromosome preparations stained with quinacrine and photographed prior to hybridisa-

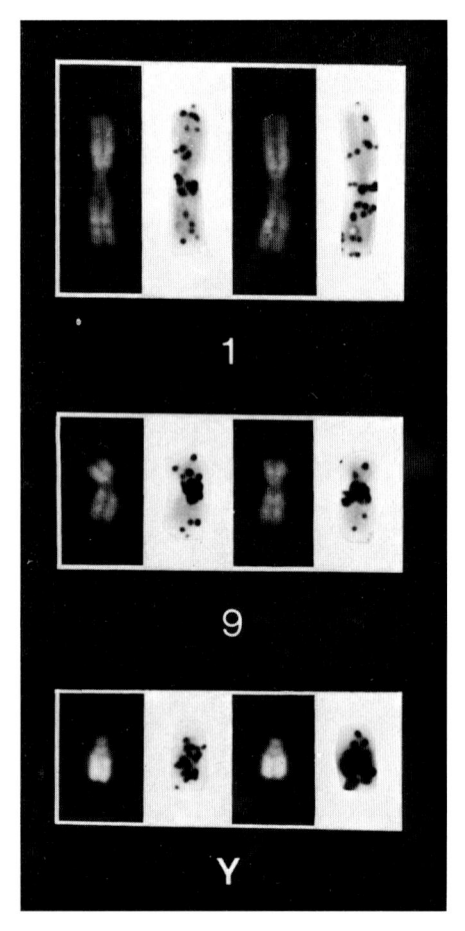

Fig. 14.1. Chromosomes 1, 9, and the Y chromosomes of a 47, XYY human male with quinacrine staining (*left*), and after in situ hybridisation with human satellite cRNA IV (*right*). Note minor site of hybridisation in the proximal region of the short arm of chromosome 1, and major sites of hybridisation in chromosome 9 and the Y. Courtesy of Professor H.J. Evans. [Evans et al., Nature (London) *251*, 346–347 (1974)]

tion, thus permitting a clear identification of each chromosome. This method was initially introduced by Evans et al. (1974) who reported that the Y chromosome in man was a major site of hybridisation with the four human satellite cRNAs, especially in the long arm distal region which stains brilliantly with quinacrine (Fig. 14.1). The same method was later used by Gosden et al. (1975) to study the chromosome distribution of the four satellite DNAs (I, II, III and IV) in man, and in this study, a quantitative analysis of grain distribution distinguished major from minor sites of hybridisation. The distribution of the four satellite DNAs reported in this work generally coincided with previous reports, although results were not strictly identical. One difference, for example, was that cRNA II was found to be the only satellite cRNA to hybridise significantly above the background in the centromere and secondary constriction region of chromosome 16.

These reports showed that the four human satellites were not distributed throughout the whole chromosome complement in man, under experimental conditions ($T = 65\,°C$) of hybridisation above the T_{OPT} for satellite I, II, and III DNA. On the other hand, most satellite-rich regions were sites of hybridisation with more than one kind of satellite cRNA, a finding that could have resulted from three possibilities. The first was contamination of satellite fractions, so that each cRNA transcribed consisted actually of more than one kind of complementary sequence. The second was cross-hybridisation which could result from sequence homology between two or more satellite DNAs, so that their transcribed cRNAs formed heterologous cRNA-DNA hybrids. The third possibility was the existence of more than one kind of satellite DNA sequence in many heterochromatic regions of the human chromosome complement.

The Distribution of Homologous Sequences to the Four Human Satellite I, II, III and IV DNA in the Chromosome Complement of the Great Apes

A greater understanding of the findings relating to man was achieved by experiments of interspecific in situ hybridisation, using human satellite cRNAs and applying them to chromosome preparations of *Pan troglodytes, Gorilla gorilla* and *Pongo pygmaeus*. This approach was initially used by Jones et al. (1972), who detected homologous sequences to human satellite III in the chromosome complement of these ape species. When applying human cRNA II to chromosomes of *Pan troglodytes,* however, no detectable hybridisation was observed. More recently, Gosden et al. (1977) and Mitchell et al. (1977) carried out in situ hybridisations using satellite cRNA, I, II, IV, and III respectively, with chromosome preparations of great ape species that had been stained with quinacrine and photographed prior to hybridisation. This permitted not only precise chromosome identification within the chromosome complement of each species, but also comparisons between interspecific homologues in relation to hybridisation. These studies showed that homologous sequences

Table 14.1. Level of hybridisation of human satellite cRNA I, II and III, and IV with human and great ape chromosomes

Species	Satellite cRNA				
	I	II	III	IV	Total
Homo	18	32	26	45	131
Pan	83	0	142	41	266
Gorilla	75	88	136	82	311
Pongo	75	67	95	5	242

Numbers indicate total grains per cell corrected for background according to Gosden et al., Chromosoma (Berlin) 63, 253–271 (1977).

Table 14.2. In situ hybridisation of human cRNA I, II, III and IV with hominoid metaphase chromosomes[a,b]

Homo	I	II	III	IV	Pan	I	III	IV	Gorilla	I	II	III	IV	Pongo	I	II	III
1	+	+	+	+	1	—	+	—	1	—	+	+	+	1	—	—	—
2p	—	—	—	—	12	—	—	—	12	+	+	+	+	12	+	+	+
2q	—	—	—	—	13	+	+	+	11	—	—	—	—	11	+	+	+
3	—	—	—	—	2	—	—	—	2	—	—	—	—	2	—	—	—
4	—	—	—	—	3	—	—	—	3	—	—	—	—	3	—	—	—
5	+	—	+	+	4	—	—	—	4	+	+	+	—	4	—	—	—
6	—	—	—	—	5	—	—	—	5	—	—	—	—	5	—	—	—
7	+	—	+	+	6	+	+	+	6	—	—	—	—	10	—	—	—
8	—	—	—	—	7	—	—	—	7	—	—	—	—	6	—	—	—
9	+	+	+	+	11	+	+	+	13	+	+	+	+	13	+	+	+
10	—	—	—	+	8	—	—	—	8	—	—	—	—	7	—	+	—
11	—	—	—	—	9	—	—	—	9	—	—	—	—	8	+	—	—
12	+	—	—	—	10	—	—	—	10	—	—	—	—	9	+	—	—
13	+	+	+	+	14	+	+	+	14	+	+	+	+	14	+	+	+
14	+	+	+	+	15	+	+	+	18	+	+	+	+	15	+	+	+
15	+	+	+	+	16	+	+	+	15	+	+	+	+	16	+	+	+
16	—	+	—	—	18	+	+	+	17	+	+	+	+	18	—	—	—
17	—	+	—	+	19	+	+	+	19	—	—	—	—	19	—	—	—
18	—	—	—	—	17	+	+	+	16	+	+	+	+	17	+	+	+
19	—	—	—	—	20	—	—	—	20	—	—	—	—	20	—	—	—
20	+	+	+	+	21	+	+	+	21	—	—	—	—	21	—	—	—
21	+	+	+	+	22	+	+	+	22	—	—	—	—	22	+	+	+
22	+	+	+	+	23	+	+	+	23	+	+	+	+	23	+	+	+
X	—	—	—	—	X	—	—	—	X	—	—	—	—	X	—	—	—
Y	+	+	+	+	Y	+	+	+	Y	+	+	+	+	Y	+	+	+

\+ Indicates detectable hybridisation above grain background.
[a] Gosden et al. (1977).   [b] Mitchell et al. (1977).

Chromosomes homologies between species follow the Paris Conference (1971); Supplement 1975 criterion and the additional changes proposed at the Stockholm Conference (1978).

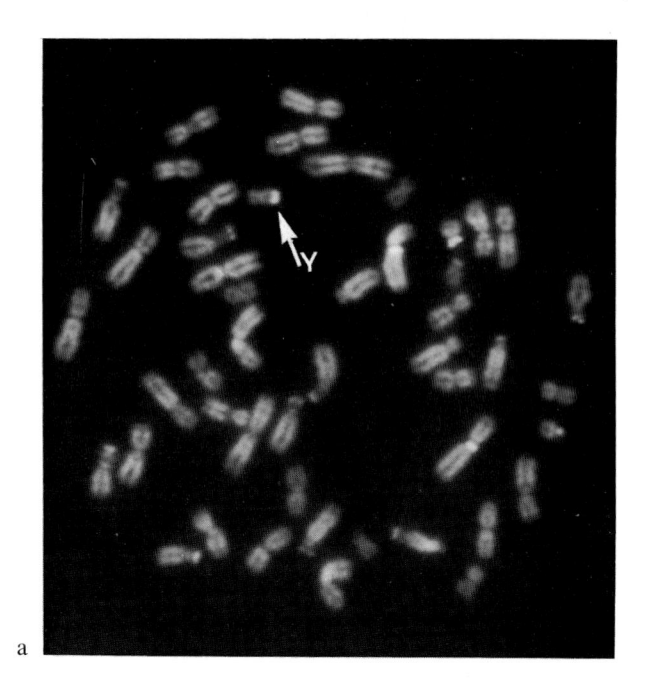

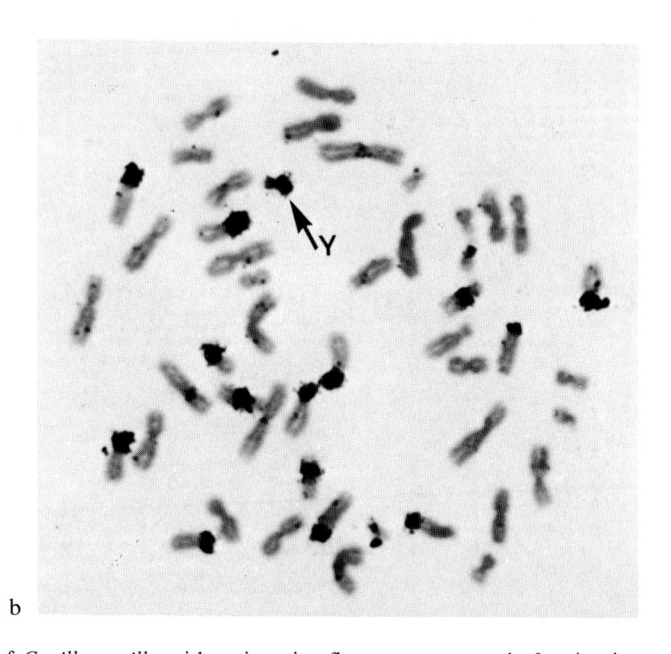

Fig. 14.2. Chromosomes of *Gorilla gorilla* with quinacrine fluorescence **a** and after in situ hybridisation with human satellite cRNA III **b**. Note high grain concentration in the Y chromosome (arrow), and some autosomes

to all four human satellite DNAs were detectable only in chromosomes of *Gorilla* among the three ape species studied. No hybridisation was detected in the chromosomes of *Pan troglodytes* with cRNA II, as previously reported, but this human cRNA hybridised with chromosomes of *Gorilla*, and also of *Pongo*, a species phylogenetically more distant from man than *Pan* and *Gorilla*. On the other hand, human satellite cRNA IV did not hybridise in significant amounts above background to any autosome in *Pongo*, a species where hybridisation with cRNA IV was only detectable at the Y chromosome. As estimated from the average grain count in each species (Table 14.1), the level of hybridisation of cRNA I and III was higher in any species of great ape than in man. In both *Gorilla* and *Pongo* the total grain count for cRNA II was approximately twice that observed in man, and was similar to the amount of cRNA I hybridised in these two ape species. Similar levels of hybridisation were observed for cRNA IV in man and *Pan troglodytes*, whereas in *Gorilla* the level was approximately twice as high.

The distribution of hybridisation of the four human satellite cRNAs in the chromosome complement of man and the three species of great apes is shown in Table 14.2. The sites where hybridisation was detected were mainly the centromere or near centromere region of the chromosomes listed in Table 14.2, the secondary constriction region of chromosome 17 and 18 in *Gorilla* (Fig. 14.2), a situation comparable to the secondary constriction region of chromosome 1, 9, and 16 in man, and, finally, the Y chromosome of all species (see also Fig. 14.3).

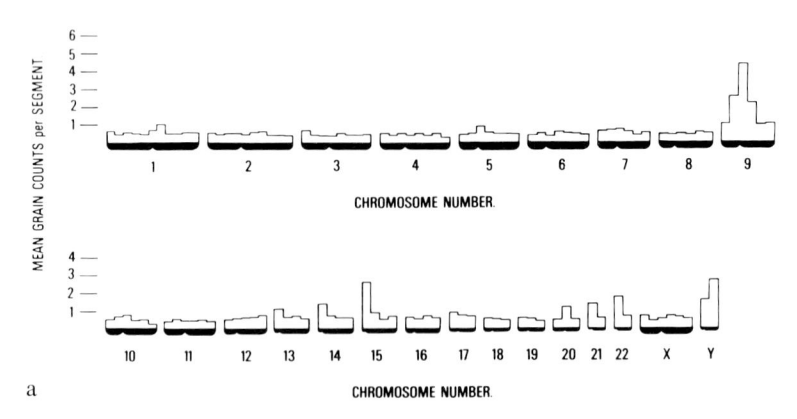

Fig. 14.a–d. In situ hybridisation of human satellite III cRNA on metaphase chromosomes. Hybridisation is shown as the mean number of grains per chromosomal segment. *a* man; *b* chimpanzee (spp. *troglodytes*); *c* Gorilla and *d* Orangutan. [Mitchell et al., Chromosoma (Berlin) *61*, 345–358 (1977), Courtesy of R.A. Mitchell]

143

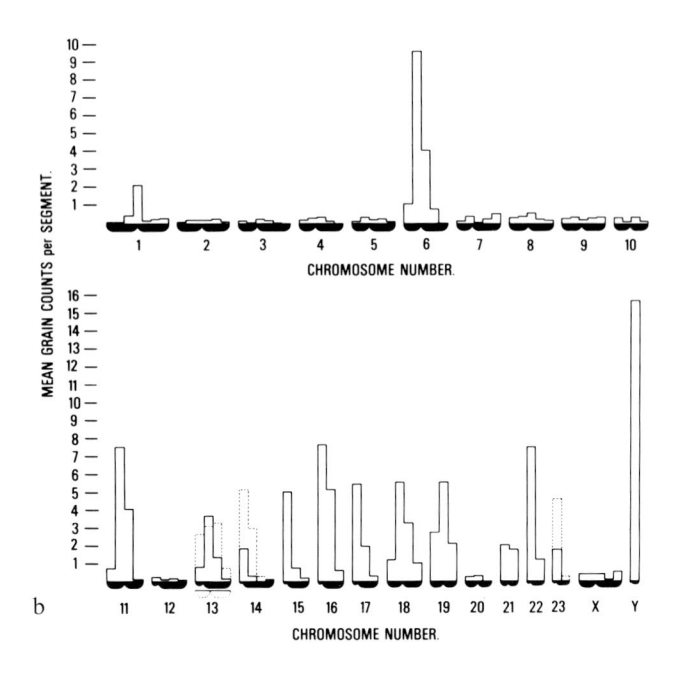

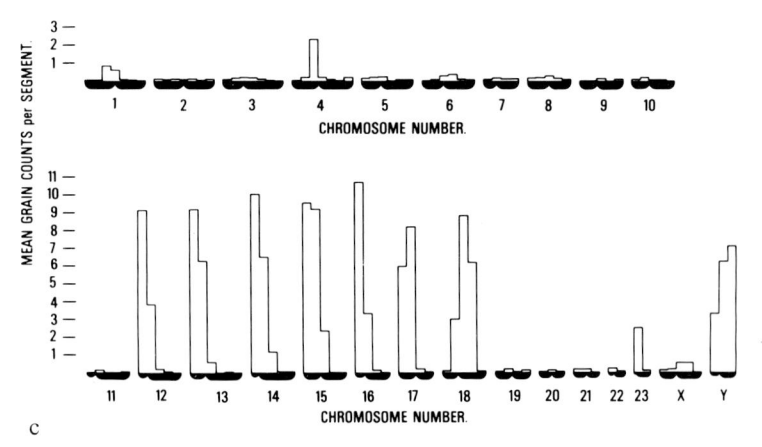

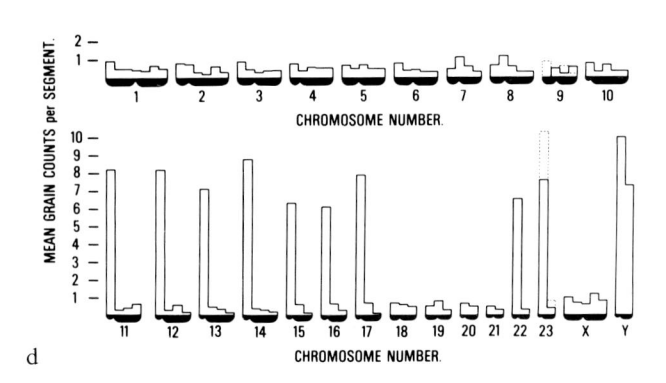

Interspecific Chromosome Homologies in the Hominidae in Relation to Hybridisation. Independent Amplification of Highly Repetitive DNAs after Speciation

The experimental conditions under which interspecific in situ hybridisation was carried out were stringent, allowing only well-matched cRNA–DNA duplexes to be formed (Gosden et al., 1977). For this reason, the differences observed in the grain counts of the different cRNAs within and between species were considered to be due to a different degree of amplification of these sequences in the chromosome complement of man and the great apes. Lack of hybridisation of cRNA II with chromosomes of *Pan troglodytes* was considered to be indicative that satellite II DNA lacked close homology with the other three satellites in contrast to previous reports of cross-hybridisation in vitro between satellite II and III (Melli et al., 1975). If such close homology existed between these human satellite DNAs, some cross-hybridisation would occur with cRNA II and chromosomes of *Pan troglodytes,* since these chromosomes hybridised with satellite cRNA III. On the other hand, the restricted hybridisation of cRNA IV to *Pongo* was also considered to result from a partial amplification of this repetitive sequence in this species. However, we could not rule out completely the possibility that this DNA sequence might have been absent from *Pongo,* and that heterologous hybridisation between cRNA IV and DNA sequences homologous to satellite I or III DNA did not occur in the Y chromosome.

Interspecific homologues were compared in relation to (1), the kind of satellite cRNA hybridised (2), the relative amounts of hybridisation detected, and (3), the site of hybridisation within chromosomes. This comparison showed, for example, that a minor site of hybridisation was detected, although with different satellite cRNAs, in the centromere region of HSA 1, PTR 1, and GGO 1 (Table 14.2). In the human chromosome, hybridisation also occurred in the secondary constriction region, which is absent in its ape homologues, but in HSA 1 is located at the proximal region of the long arm. In PTR 1 and GGO 1 hybridisation occurred at the centromere region, and also at the proximal region of the short arm, thus at the same arm as in man, although at a region which is not heterochromatic. Chromosome 2 in man, for example, did not show hybridisation with any satellite cRNA, but its corresponding arm homologues in *Pongo* hybridised with cRNA I, II, and III. In *Pan troglodytes,* chromosome 13 (HSA 2q) had a major site of hybridisation with cRNA I, III, and IV, but PTR 12 (HSA 2p) did not bind any cRNA.

In *Gorilla*, the inverse situation was observed: GGO 12 (HSA 2p) hybridised with the four cRNAs, but GGO 11 (HSA 2q) did not (Table 14.2). Since chromosome 2 in man has probably arisen from a telomeric fusion of its respective "arm homologues", the absence of satellite DNA in HSA 2 is unlikely to have resulted from loss of repetitive DNA due to chromosome rearrangement, as we would expect it to have occurred if two acrocentric chromosomes fused by their centromere region. Chromosome 7 in man shows but a minor site of hybridisation, but in its homologue in *Pan troglodytes* (PTR 6), the subcentromeric region of this chromosome is a major site of hybridisation with cRNA I, III and IV (see also Fig. 15.1 and 15.4). Chromosome 16 in man only binds cRNA II in the centromere and secondary constriction region, but its homologue in *Pan troglodytes* (PTR 18) binds cRNA I and III, its homologue in Gorilla (GGO 17) the four cRNAs, and its homologue in *Pongo* (PPY 18) none (Table 14.2). Here again, differences between interspecific homologues could not be explained by chromosome rearrangement.

These results indicated that chromosomes recognized as interspecific homologues had different kinds and amounts of sequences homologous to the four human satellites. Considering that interspecific homologies inferred from chromosome morphology and banding are generally confirmed by comparative gene mapping of single gene sequences (Chap. 12), the finding of different kinds of repetitive DNAs must then have resulted from their being amplified independently after speciation. If amplification had occurred before speciation, each species would have inherited a somewhat similar distribution and interspecific homologues would contain similar kinds of repetitive sequences in the same way as they contain similar structural genes. If amplification had occurred before speciation, we would have found a situation comparable to that of the 5S rDNA genes which are located at the same chromosome region of the same chromosome over a wide range of primate species (Chap. 16).

A further indication of independent amplification after speciation is the fact that satellite cRNA II did not hybridise with any chromosomes in *Pan troglodytes,* but hybridised with chromosomes of *Gorilla gorilla* and *Pongo pygmaeus.* This makes it highly unlikely that satellite II could have originated in the lineage of man after the man-chimpanzee divergence, as suggested by Jones et al. (1972). On the contrary, assuming that no satellite loss occurred in *Pan troglodytes,* the original sequence must have been present in few or single copies in the common ancestor of the Hominidae. Otherwise, if amplification had occurred before speciation, one kind of sequence (satellite II) must have been lost in *Pan troglodytes,* and probably another (satellite IV) in the autosomes of *Pongo pygmaeus*. But, even assuming satellite loss, the difference observed between interspecific homologue chromosomes could still not be explained (Seuánez et al., 1977).

As a conclusion, these experiments indicate that homologous sequences to the four human satellites have been independently amplified in the Hominidae, in a way comparable to the α-satellite of the guinea pig, the kangaroo rat, the antelope squirrel, and the pocket gopher (Chap. 13). However, conclusive evidence in favour or against this hypothesis should be provided when the sequence of each of the four human satellites becomes known, and studies can be based on more precise techniques than in situ hybridisation.

References

Evans, H.J., Gosden, J.R., Mitchell, A.R., Buckland, R.A.: Location of human satellite DNAs on the Y chromosome. Nature (London) *251,* 346–347 (1974)

Gall, J., Pardue, M.L.: Formation and detection of RNA-DNA hybrid molecules in cytological preparations. Proc. Nat. Acad. Sci. USA *63,* 378–383 (1969)

Gosden, J.R., Mitchell, A.R., Buckland, R.A., Clayton, R.P., Evans, H.J.: The location of four human satellite DNAs on human chromosomes. Exp. Cell Res. *92,* 148–158 (1975)

Gosden, J.R., Mitchell, R.A., Seuánez, H.N., Gosden, C.: The distribution of sequences complementary to satellite I, II and IV DNAs in the chromosomes of the chimpanzee *(Pan troglodytes),* gorilla *(Gorilla gorilla),* and orangutan *(Pongo pygmaeus).* Chromosoma (Berlin) *63,* 253–271 (1977)

Jones, K.W., Corneo, G.: Location of satellite and homogeneous DNA sequences on human chromosomes. Nature New Biol. *233,* 267–271 (1971)

Jones, K.W., Prosser, J., Corneo, G., Ginelli, E.: The chromosomal location of human satellite III. Chromosoma (Berlin) *42,* 445–451 (1973)

Jones, K.W., Prosser, J., Corneo, G., Ginelli, E., Bobrow, M.: Satellite DNA constitutive heterochromatin and human evolution. In: Modern aspects of cytogenetics: constitutive heterochromatin in man. Symposia Medica Hoechst. Pfeiffer, R.A. (ed.), Vol. VI, pp. 45–61. Stuttgart, New York 1972

Jones, K.W., Purdom, I.F., Prosser, J., Corneo, G.: The chromosomal localization of human satellite DNA I. Chromosoma (Berlin) *49,* 161–171 (1974)

Melli, M., Ginelli, E., Corneo, G., di Lernia, R.: Clustering of the DNA sequences complementary to repetitive nuclear DNA of HeLa cells. J. Mol. Biol. *93,* 23–38 (1975)

Mitchell, A.R., Seuánez, H.N., Lawrie, S., Martin, D.E., Gosden, J.R.: The location of DNA homologous to human satellite III DNA in the chromosomes of chimpanzee *(Pan troglodytes),* gorilla *(Gorilla gorilla)* and orangutan *(Pongo pygmaeus).* Chromosoma (Berlin) *61,* 345–358 (1977)

Moar, M.H., Purdom, I.F., Jones, K.W.: Influences of temperature on the detectability and chromosomal distribution of specific DNA sequences by in situ hybridisation. Chromosoma (Berlin) *53,* 345–359 (1975)

Seuánez, H., Mitchell, A., Gosden, J.R.: Constitutive heterochromatin in the Hominidae in relation to four satellite DNAs in man and homologous sequences in the great apes. In: Proc. III Latin-Amer. Congr. Genet, Montevideo, 1977, Joint Seminar and Workshop. Aspects of Chromosome Organization and Function. Drets, M.E., Brum-Zorrilla, N., Folke, G.A. (eds.), 171–178 (1977)

Chapter 15 DNA Composition of Constitutive Heterochromatin in the Chromosome Complement of Man and the Great Apes

Constitutive Heterochromatin as Demonstrated by C-Banding

In the previous chapter, we have shown how human satellite cRNA I, II, III, and IV hybridised with human and ape chromosomes mainly in regions of constitutive heterochromatin. In the great apes, as in man, it was also observed that many sites of hybridisation were the same for more than one satellite cRNA. In view of the lack of hybridisation of human cRNA II with chromosomes of *Pan troglodytes,* of human cRNA IV with chromosomes of *Pongo,* and, thirdly, of the different satellite cRNAs hybridised with interspecific homologue chromosomes, it was suggested that each sequence hybridising at each site was actually different from, and lacked close homology with, one another (Gosden et al., 1977). Based on this assumption it was also suggested that man and the species of great apes contained different kinds of highly repetitive DNA sequences arranged in tandem blocks at the same chromosomal site, a situation comparable to that observed in *Drosophila melanogaster* by Peacock et al. (1973).

When the distribution and the level of hybridisation within each species was compared with those of other species, some interesting findings were observed. In man, for example, not all heterochromatic regions were sites of hybridisation, but the level of hybridisation was positively correlated with the size of the heterochromatic region as demonstrated with C-banding. This was the case of the Y chromosomes long-arm distal region, and of the secondary constriction region of chromosome 9, both of them accounting for the two major sites of hybridisation of the human chromosome complement. On the other hand, some other heterochromatic regions were only minor sites of hybridisation, whereas others showed no detectable hybridisation under the same experimental conditions. This unequal distribution of highly repetitive DNA in the human chromosome complement differed completely from that found in other organisms such as the mouse, for example, in which a roughly similar concentration of grains was observed in every chromosome, except for the Y chromosome in which no hybridisation was detected with mouse satellite cRNA (Pardue and Gall, 1970). In the great apes, as in man,

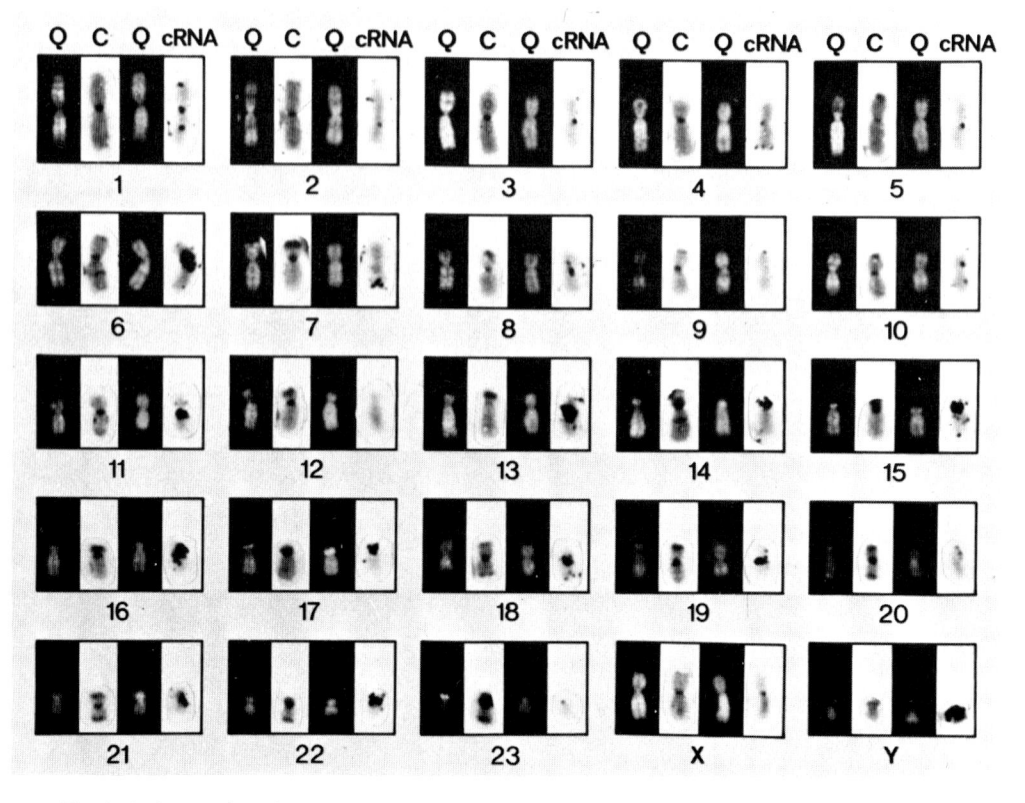

Fig. 15.1. Comparison between C-band regions (*left*), and sites of hybridisation with human satellite cRNA III (*right*) in *Pan troglodytes*. Each chromosome had been previously identified with Q-banding

the distribution of hybridisation of the four human satellite cRNAs was also unequal, and many heterochromatic regions such as those of the centromere of the large metacentric chromosomes, did not show detectable hybridisation (Figs. 15.1, 15.2, and 15.3). When sites of hybridisation were compared to C-band regions in the chromosome complement of the great apes, a positive correlation was also found, in some cases, between the size of a C-band region and the level of hybridisation, or grain concentration. This was the case of the secondary constriction region of chromosome 17 and 18 in *Gorilla,* for example. Also, in the Y chromosome of *Gorilla* and *Pongo* large C-band regions coincided with a high level of hybridisation (Fig. 15.4). However, detectable amounts of hybridisation were observed in regions that could not be demonstrated as constitutive heterochromatin by C-banding, like, for example, the proximal region of the short-arm of chromosome 1 in *Pan troglodytes* and in *Gorilla*.

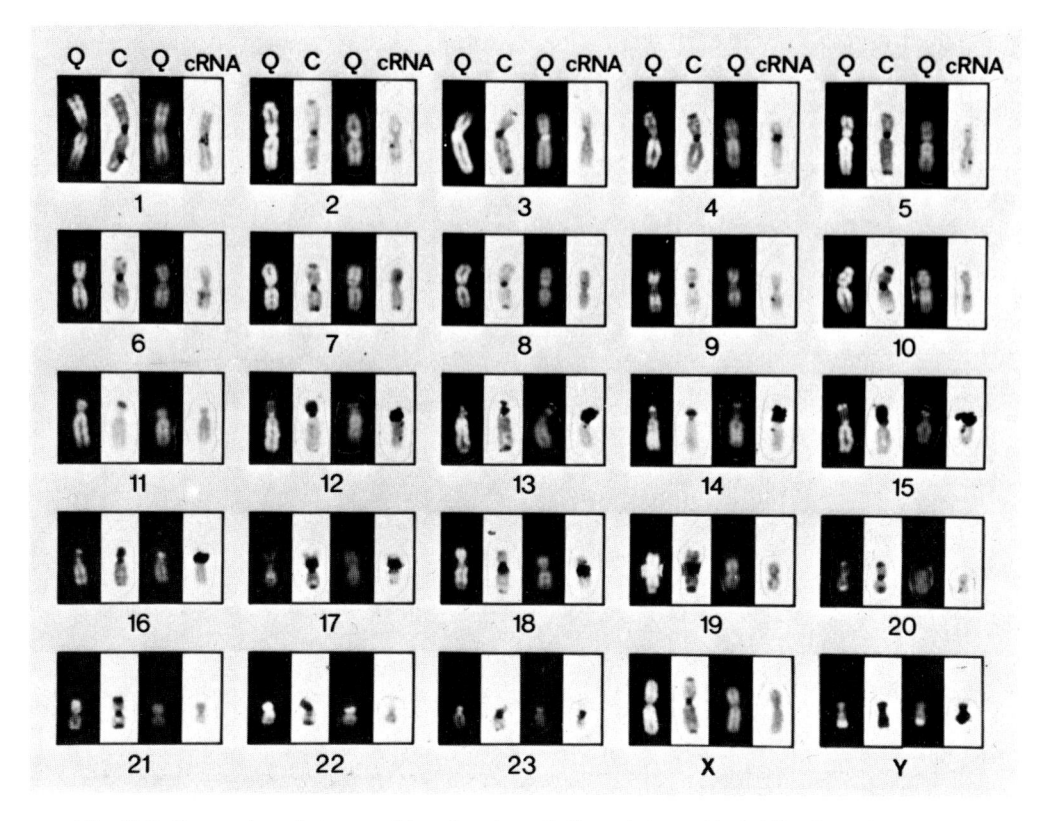

Fig. 15.2. Comparison between C-band regions (*left*), and sites of hybridisation with human satellite cRNA III (*right*) in *Gorilla gorilla*. Each chromosome had been previously identified with Q-banding

In this region there is no heterochromatic secondary constriction as there is in the proximal region of the long-arm of chromosome 1 in man (Fig. 15.4). An even more striking example was chromosome 6 in *Pan troglodytes* (PTR 6), where there is a small heterochromatic region in the centromere as there is in its human homologue (HSA 7). However, PTR 6 was a major site of hybridisation with cRNA I, III and IV; the distribution of grains within this chromosome covering not only the centromere, but extending to the adjacent non-heterochromatic regions of the short and long-arm (Fig. 15.4). Actually, the grain concentration found in this chromosome was higher than that found in chromosome 9 of man with satellite cRNA I, III, and IV (see Fig. 14.3a and b for a comparison of hybridisation with cRNA III). Furthermore, the Y chromosome in *Pan troglodytes* was also a major site of hybridisation with cRNA I, III, and IV; in all cases the entire Y chromosome was

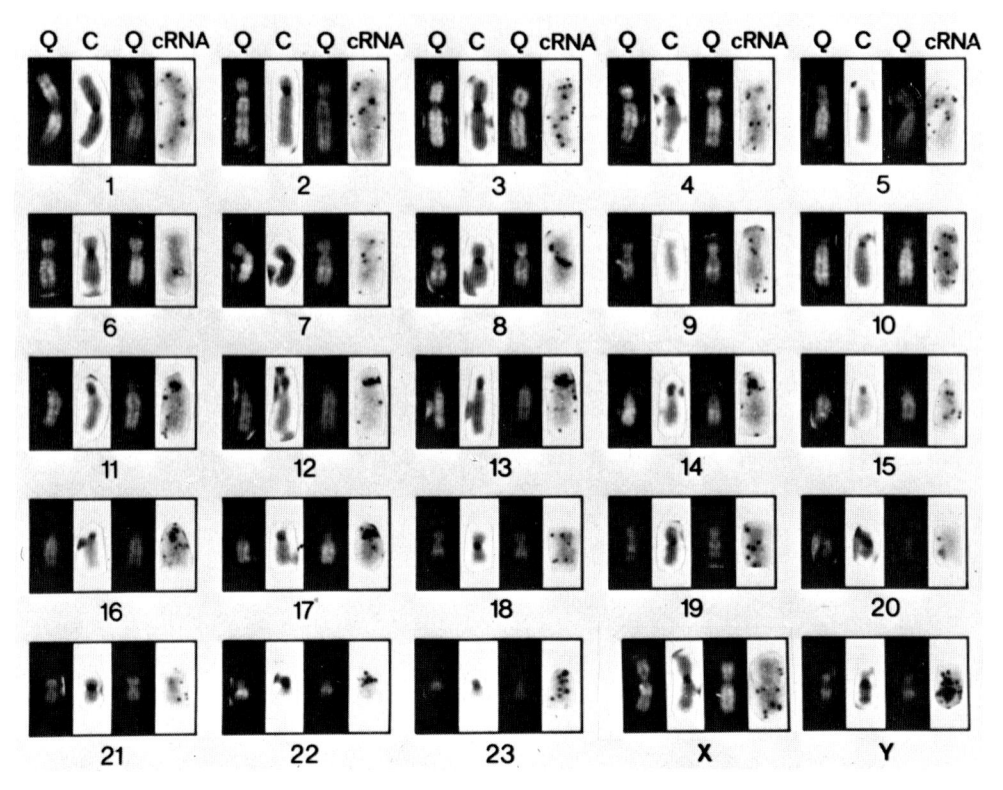

Fig. 15.3. Comparison between C-band regions (*left*), and sites of hybridisation with human satellite cRNA III in *Pongo pygmaeus*. Each chromosome had been previously identified with Q-banding

obliterated with grains, as shown in Fig. 15.5, although only its very small short arm is positively C-banded. Here again, this chromosome contained a higher amount of sequences homologous to human satellite I, III, and IV than the human Y chromosome, where there is a large heterochromatic region in the long arm. These examples demonstrated that C-band size in the Hominidae are not necessarily correlated to the amount of highly repetitive DNA sequences, contrary to what was previously observed in the human chromosome complement.

A detailed comparison between the sites of hybridisation with constitutive heterochromatin region in chromosomes of the great apes showed that the interstitial C-band region of chromosome 6 and 14 in *Pan troglodytes* did not correspond to sites of hybridisation with any human satellite cRNA (Fig. 15.4). A similar finding was observed in all terminal heterochromatic regions in *Pan troglodytes* and *Gorilla;* none of these was a site of hybridisation with any of the four human satellite cRNAs,

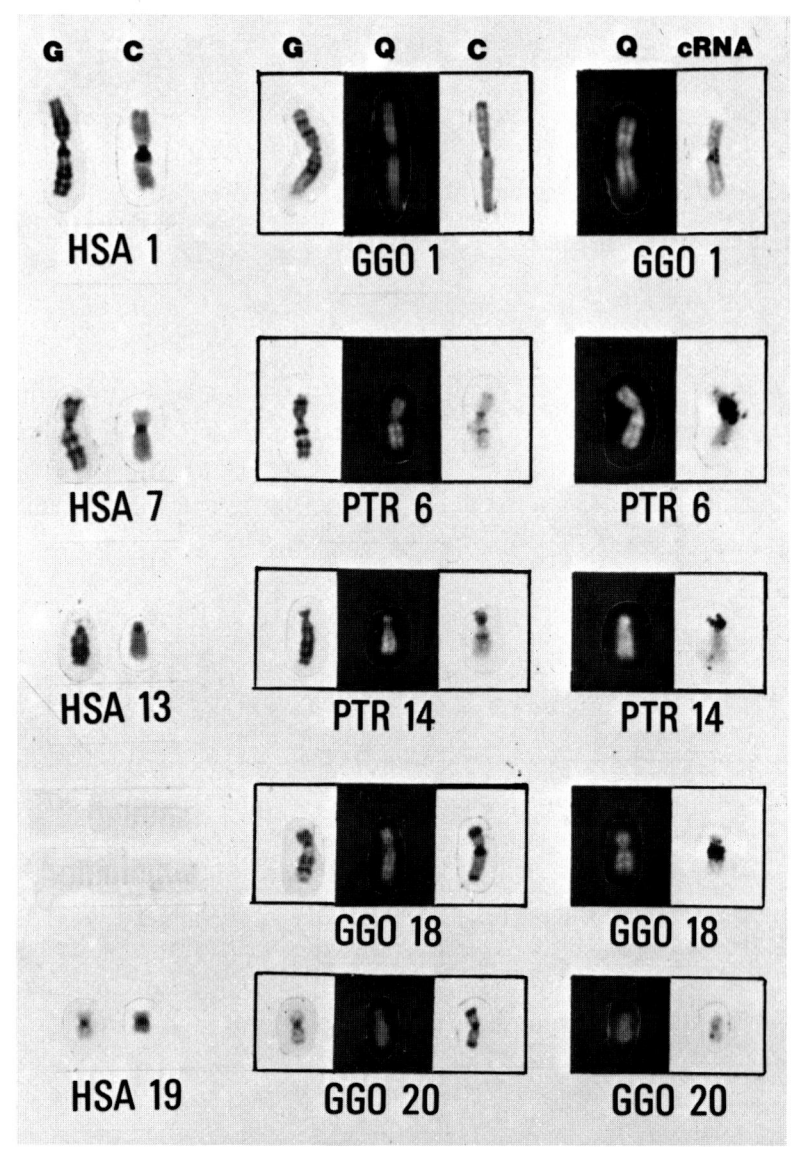

Fig. 15.4. Chromosomes of man and the great apes compared with G-, Q-, and C-banding. Compare the positively C-banded regions of these chromosomes (*central column, right*) with the sites of hybridisation in the same chromosomes (*right column*)

although some of these telomeric heterochromatic regions could sometimes be of considerable size (see Fig. 6.6).

These results must be carefully evaluated, since there seems to be no one to one specificity between C-band regions and sites of hybridisation with the four human satellite cRNAs (Seuánez et al., 1977). One of the

151

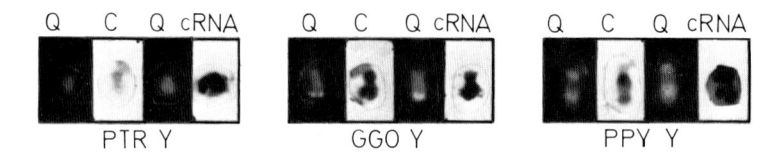

Fig. 15.5. The Y chromosome of *Pan troglodytes* (PTR), *Gorilla gorilla* (GGO) and *Pongo pygmaeus* (PPY) with Q- and C-banding, and with Q-banding prior to in situ hybridisation with human satellite cRNA III. Note the small C-band region in PTR Y

reasons for this discrepancy might be due to the existence of highly repetitive DNA sequences in a positively C-banded region, in amounts below those detectable by in situ hybridisation under stringent experimental conditions. Another possibility is that C-band regions not showing detectable hybridisation actually contain very different highly repetitive DNA sequences from the four human satellite DNAs, so that cross-reaction between the former and the latter does not occur. This may well explain the lack of hybridisation observed between any of the four human satellite cRNAs with the terminal heterochromatic regions in *Pan troglodytes* and *Gorilla,* and also with the interstitial heterochromatic regions in *Pan troglodytes*. Finally, the possibility that C-band regions may not contain highly repetitive DNA should not be overlooked. In the Chinese hamster, for example, large heterochromatic regions exist in the sex chromosomes, although they do not contain significant amounts of highly repetitive DNA (Arrighi et al., 1974). This finding suggests that, in this species, C-banding must probably result from DNA packaging, only indirectly from DNA composition; the structural organization of the DNA in the sex chromosomes must probably depend on the nature and composition of chromosome proteins. Moreover, it is important to mention that, in other organisms, C-bands have been found to contain entirely different kinds of DNA. For example in the mouse, C-banding reflects A-T-rich satellite sequences (Pardue and Gall, 1970); in the oxen, G-C-rich satellite sequences (Kurnit et al., 1973); while in *Microtus agrestis,* only moderately repetitive DNA (Arrighi et al., 1970).

For all the above-mentioned reasons, the appearance of heterochromatic regions in the phylogeny of the human or great ape chromosomes cannot be explained merely as having resulted from the amplification of a simple sequence in a chromosome region. As we have previously pointed out, there are good examples of extensive amplification which have not resulted in large heterochromatic blocks or secondary constrictions, as we might have expected. The appearance of C-band regions must thus have necessarily demanded the acquisition of a new level of structural organization of DNA, most probably resulting from evolutionary change in chromosomes at the protein level.

G-11 Regions and Satellite III-Rich Regions

Other techniques have been used to demonstrate selectively some regions of the constitutive heterochromatin of the human chromosome complement, such as the G-11 technique (Bobrow et al., 1972; Gagné and Laberge, 1972). These techniques have also been used to study the chromosome complement of *Pan troglodytes* (Bobrow and Madan, 1973; Lejeune et al., 1973); *Gorilla gorilla* (Bobrow et al., 1972; Dutrillaux et al., 1973; Pearson, 1973); and *Pongo pygmaeus* (Dutrillaux et al., 1975). Each of these reports used a different criterion of nomenclature and of presumptive homologies between human and ape chromosomes, but a comparative table of results can be obtained if their observations are adapted to the standard criterion of the Paris Conference (1971) supplement (1975); Table 15.1. In this table we have also added the sites of hybridisation with cRNA III in the chromosome of these species as presented by Mitchell et al. (1977), since it is claimed (see Bobrow et al., 1972; Bobrow and Madan, 1973; Pearson, 1973; Jones, 1976) that the G-11 technique has selectively demonstrated satellite III DNA rich regions in the chromosome complement of man and the great apes.

From data in Table 15.1 it can be seen that in man, the location of G-11 regions coincides with satellite III-rich regions except for Nos. 4, 5, 7, 10, and 17. In *Pan troglodytes* both reports on the distribution of G-11 regions are basically coincident (Bobrow and Madan, 1973; Lejeune et al., 1973), except that in the report of Lejeune et al. (1973) it is the proximal region of the long-arm of chromosome 12 (HSA 2p) that shows positive G-11 staining, and in that of Bobrow et al. (1973) it is the proximal region of chromosome 13 (HSA 2q) that is stained. A comparison with the sites of in situ hybridisation with cRNA III shows that there is good correlation between these sites and the G-11 stained regions, except for chromosome 12 in which there is no satellite III. The Y chromosome is stained only in its very short arm, but it is completely obliterated when hybridised with cRNA III. The pattern of G-11 banding in *Gorilla gorilla* has been shown in photographs by Bobrow et al. (1972) and Pearson (1973), but with the exception of the Y, the chromosomes were not identified. Dutrillaux et al. (1973) studied the Giemsa 11 pattern of this species

Table 15.1. Comparison of the G-11 staining and in situ hybridisation of human cRNA III to hominoid metaphase chromosomes

Homo			Pan troglodytes				Gorilla gorilla			Pongo pygmaeus		
Chr.	cRNA III	G-11[a]	Chr.	cRNA III	G-11[a]	G-11[b]	Chr.	cRNA III	G-11[c]	Chr.	cRNA III	G-11[d]
1	q	q	1	p	p	cen	1	p	cen	1	—	cen?
2p	—	—	12	—	—	q	12	p	p	12	p	p
2q	—	—	13	q	q	—	11	—	—	11	p	p
3	—	—	2	—	—	—	2	—	—	2	—	—
4	—	q	3	—	—	—	3	—	—	3	—	—
5	q	—	4	—	—	—	4	q	—	4	—	—
6	—	—	5	—	—	—	5	—	—	5	—	—
7	—	q	6	q	p	p	6	—	p	10	—	cen
8	—	—	7	—	—	—	7	—	—	6	—	—
9	q	q	11	q	q	q	13	p	p	13	p	p
10	—	q	8	—	—	—	8	—	p	7	—	q
11	—	—	9	—	—	—	9	—	—	8	—	—
12	—	—	10	—	—	—	10	—	—	9	—	—
13	p	p	14	p	p	p	14	p	p	14	p	p
14	p	p	15	p	p	p	18	q	q	15	p	p
15	p	p	16	p	p	p	15	p	p	16	p	p
16	—	—	18	p	p	p	17	p	p	18	—	—
17	—	p	19	q	q	cen	19	—	—	19	—	p
18	—	—	17	p	—	p	16	p	p	17	p	p
19	—	—	20	—	—	—	20	—	q	20	—	—
20	cen	p	21	q	q	q	21	—	—	21	—	—
21	p	p	22	p	p	p	22	—	p	22	p	p
22	p	p	23	p	p	p	23	p	p	23	p	p
X	—	—	X	—	—	—	X	—	—	X	—	—
Y	q	q	Y	p+q	p	?	Y	p+q	q[e]	Y	q	?

p = Short arm of chromosome.
q = Long arm of chromosome.
cen = Centromere.
? = Staining not clearly visible.

[a] Bobrow and Madan (1973).
[b] Lejeune et al. (1973).
[c] Dutrillaux et al. (1973).
[d] Dutrillaux et al. (1975).
[e] Bobrow et al. (1972).

in a female animal in which all chromosomes were identified, and their findings are presented in Table 15.1. It is important to note, however, that the results shown by Bobrow et al. (1972), and Pearson (1973), differ from those of Dutrillaux et al. (1973). In the former reports there are three large metacentric chromosomes with their entire short-arm region intensely stained with Giemsa, and this was interpreted as one autosomal pair and probably the X chromosome (Bobrow et al., 1972), whereas in the latter report the size of G-11 regions was small and never involved whole arms of metacentric chromosomes or the X chromosome. There is no explanation of this apparent difference, and the fact that G-11 was obtained in chromosomes by different procedures is an unlikely explanation. The sites of G-11 staining shown by Dutrillaux et al. (1973) and the sites of hybridisation with cRNA III are generally coincident except for Nos. 4, 6, 8, 20, and 22. In comparison, results shown by Bobrow et al. (1972) and Pearson (1973) coincide for the Y chromosome, but hybridisation with cRNA III never covers whole arms of large metacentric chromosomes or of the X chromosome.

The G-11 pattern of *Pongo pygmaeus* presented by Dutrillaux et al. (1975) generally agrees except for three chromosome regions. The centromeric region of chromosome 10, the subcentromeric region of chromosome 7 and region above the centromere of chromosome 19 all show a positive G-11 region, but no satellite III. This comparison suggests that although the sites of G-11 staining and the satellite III-rich regions generally coincide, there is no one-to-one specificity between both events, since it is possible to find regions of positive staining with no satellite III, and vice versa. One of the ways to clarify this point completely would be to do G-11 staining and in situ hybridisation on the chromosome preparations obtained from the same individual, either human or ape, so that we can discount the possibility that variations in the amount and location of satellite DNA III between individuals might produce misleading comparisons. However, a comparison between the G-11 patterns of *Gorilla gorilla* as shown by Bobrow et al. (1972) with those presented by Dutrillaux et al. (1973) makes it unlikely that such differences in staining might correspond to differences in the location (or amount) of satellite III, between the animals studied. It rather points to the fact that the technique is less accurate in showing a constant pattern, at least in this species, and that its demonstration of satellite III-rich regions is also less precise than with in situ hybridisation techniques.

References

Arrighi, F. E., Hsu, T. C., Pathak, S., Sawada, H.: The sex chromosomes of the Chinese hamster: constitutive heterochromatin deficient in repetitive DNA sequences. Cytogenet. Cell Genet. *13*, 268–274 (1974)

Arrighi, F. E., Hsu, T. C., Saunders, P., Saunders, G.: Localization of repetitive DNA in the chromosomes of *Microtus agrestis* by means of in situ hybridisation. Chromosoma (Berlin) *32*, 224–236 (1970)

Bobrow, M., Madan, K.: A comparison of chimpanzee and human chromosomes using the giemsa 11 and other chromosome banding techniques. Cytogenet. Cell Genet. *12*, 107–116 (1973)

Bobrow, M., Madan, K., Pearson, P.: Staining of some specific regions of human chromosomes, particularly the secondary constriction of No. 9. Nature New Biol. *238*, 122–124 (1972)

Dutrillaux, B., Rethoré, M.O., Lejeune, J.: Comparaison du caryotpye de l'orangutan *(Pongo pygmaeus)* a celui de l'homme du chimpanze et du gorille. Ann. Genet. *18*, 153–161 (1975)

Dutrillaux, B., Rethoré, M.O., Prieur, M., Lejeune, J.: Analyse de la structure fine des chromosomes du gorilla *(Gorilla gorilla)*. Comparaison avec *Homo sapiens* et *Pan troglodytes*. Humangenetik *20*, 343–354 (1973)

Gagné, R., Laberge, C.: Specific cytological recognition of the heterochromatic segment of number 9 chromosome in man. Exp. Cell. Res. *73*, 239–242 (1972)

Gosden, J.R., Mitchell, R.A., Seuánez, H.N., Gosden, C.: The distribution of sequences complimentary to satellite I, II and IV DNAs in the chromosomes of the chimpanzee *(Pan troglodytes)*, gorilla *(Gorilla gorilla)*, and orangutan *(Pongo pygmaeus)*. Chromosoma (Berlin) *63*, 253–271 (1977)

Jones, K.W.: Repetitive DNA sequences in animals. Proceeding of the Leiden Chromosome Conference (1974). In: Chromosomes today. Pearson, P.L., Lewis, K.R. (eds.), Vol. 5, pp. 305–313. New York: John Wiley & Sons. Jerusalem: Israel Univ. Press 1976

Kurnit, D.M., Shafit, B.R., Maio, J.J.: Multiple satellite deoxyribonucleic acids in the calf and their relation to the sex chromosomes. J. Mol. Biol. *81*, 273–284 (1973)

Lejeune, J., Dutrillaux, B., Rethoré, M.O., Prieur, M.: Comparaison de la structure fine des chromatides d'*Homo sapiens* et de *Pan troglodytes*. Chromosoma (Berlin) *43*, 423–444 (1973)

Mitchell, A.R., Seuánez, H.N., Lawrie, S., Martin, D.E., Gosden, J.R.: The location of DNA homologous to human satellite III DNA in the chromosomes of chimpanzee *(Pan troglodytes)*, gorilla *(Gorilla gorilla)* and orangutan *(Pongo pygmaeus)*. Chromosoma (Berlin) *61*, 345–358 (1977)

Pardue, M.L., Gall, J.G.: Chromosomal localization of mouse satellite DNA. Science *168*, 1356–1358 (1970)

Paris Conference (1971); supplement 1975 – Standardisation in human cytogenetics birth defects: original article series, XI; 9, New York: National Foundation 1975

Peacock, W.J., Brutlag, D., Goldring, E., Appels, R., Hinton, C.W., Lindsley, D.L.: The organisation of repeated DNA sequences in *Drosophila melanogaster* chromosomes. Cold Spring Harbor Symp. Quant. Biol. *38*, 405–416 (1973)

Pearson, P.L.: The uniqueness of the human karyotype. In: Chromosome identification techniques and applications in biology and medicine. Caspersson, T., Zech, L. (eds.), pp. 145–151. New York, London: Academic Press 1973

Seuánez, H., Mitchell, A., Gosden, J.R.: Constitutive heterochromatin in the Hominidae in relation to four satellite DNAs in man and homologous sequences in the great apes. In: Proc. III Latin-Amer. Congr. Genet, Montevideo, 1977. Joint Seminar and Workshop. Aspects of Chromosome Organization and Function. Drets, M.E., Brum-Zorrilla, N., Folle, G.A. (eds.), 171–178 (1977)

Chapter 16 The Chromosomal Distribution of Ribosomal Genes
in Man and the Great Apes

rDNA Genes in Man

Genes coding for ribosomal RNA in eukaryotes comprise three different subunits of remarkable sequence conservation which are designated 5S, 18S, and 28S and defined in terms of size and sedimentation properties. Of these three, the 18S, and the 28S are part of a 45S sequence in association with a "spacer" DNA sequence which, unlike those coding for ribosomal RNA, has undergone substantial evolutionary change showing wide differences between species (Birnstiel and Grunstein, 1972). These sequences exist in multiple copies in the eukaryotic genome; in man it has been estimated, by competition experiments on HeLa cell DNA hybridised to labelled RNA, that there are at least 600 copies of the 18S and 28S cistrons in diploid genome; and in *Xenopus* these repetitive sequences can be isolated from main band DNA by isopycnic centrifugation as a distinct G-C rich satellite fraction, a situation that does not occur in man. The 5S rDNA sequence is also repetitive; it has been estimated that there are approximately 7600 copies per cell in HeLa cells (Hatlen and Attardi, 1971).

The distribution and location of the rDNA sequences in the human chromosome complement has been successfully accomplished by techniques of in situ hybridisation. Henderson et al. (1972) and Evans et al. (1974) have demonstrated that the 18S and 28S sequences are located at the secondary constriction region of the five pairs of acrocentric chromosomes of the human chromosome complement which comprise the D- and the G-groups. These chromosomes are frequently involved in satellite association, and in mitotic preparations they appear as clusters of rosette configurations. Both in mitotic and in meiotic preparations the secondary constriction regions of the human acrocentric chromosomes have been found to be associated with nucleoli (Ferguson-Smith and Handmaker, 1961) and are therefore referred to as nucleolar organizer regions or NORs. These regions can be specifically stained with ammoniacal silver (Ag-AS) (Goodpasture and Bloom, 1975). The Ag-AS technique probably stains protein material, but nevertheless it has been found to be a reliable procedure for demonstrating functional NORs in a large number of species, in which the distribution of silver staining has been compared to the distribution of the 18S and 28S rDNA cistrons as revealed by in situ hybridisation (Hsu et al., 1975). It must be pointed out, however, that the Ag-AS technique shows only active sites of rDNA transcription, but not necessarily all sites that contain 18S and 28S rDNA sequences. Eliceiri and Green (1969) found that in human–mouse hybrid cells, human rDNA genes were suppressed, and D. A. Miller et al. (1976) showed that in human–mouse hybrid cells in which there had been loss of human chromosomes, there was no Ag-AS staining of the human acrocentric chromosomes, although there had been in the human parental cell line from which the hybrid cell line was derived. The inverse situation was also found by O. J. Miller et al. (1976) in human–mouse hybrid lines in which there had been loss of mouse chromosomes. Even in cases where all mouse chromosomes carrying a NOR had been conserved in the hybrid, none of them was stained with the Ag-AS technique, although these chromosomes were positively stained in the parental mouse cell line. The results were interpreted as resulting from the suppression of the transcriptional activity of 18S and 28S rDNA genes in one chromosome set as a result of partial chromosome loss of the same set in the hybrid, but was not due to loss of NORs (see also Croce et al., 1977).

It has been demonstrated by in situ hybridisation that in man (Evans et al., 1974), the Rhesus monkey (Henderson et al., 1974a), the gibbon

Table 16.1. Chromosome distribution of 18S and 28S rDNA genes in the Hominidae

Man[a]	Chimpanzee[b]	Pygmy chimpanzee[b]	Gorilla[b]	Orangutan[c]
2p	12	12	12	12
2q	13	13	11	11
9	11	11	13	13
13	14	14	14	14
14	15	15	18	15
15	16	16	15	16
18	17	17	16	17
21	22	22	22	22
22	23	23	23	23

[a] Henderson et al., 1972.
[b] Henderson et al., 1976a.
[c] Gosden et al., 1978

Chromosomes containing rDNA genes in each species are shown inside squares.

(Warburton et al., 1975), the chimpanzee (Henderson et al., 1974b), the pygmy chimpanzee, and the orangutan (Henderson et al., 1976a) there are differences in the gene multiplicity among NORs. Such differences are especially clear in man where variant chromosomes with large NORs or with large terminal satellite regions have been studied (Evans et al., 1974). The amount of hybridisation of 18S and 28S RNA was significantly greater in variant chromosomes with large NORs than in those with normal-sized NORs, whereas the size of the terminal satellites was found to be unrelated to the amount of hybridisation. These results point to the fact that 18S and 28S rDNA sequences are located at the satellite stalks of the five human acrocentric chromosomes but not at their terminal satellites. Warburton et al. (1976) has demonstrated that the amount of rDNA gene copies in a chromosome is positively correlated with its frequency of satellite association in man.

18S and 28S Cistrons in the Great Apes and Other Primates

The distribution of these sequences in the chromosome complement of the great apes has been reported (Henderson et al., 1974b; 1976a; Gosden et al., 1978) with techniques of in situ hybridisation, and by Tantravahi et al. (1976) with silver staining. A comparative table of results is shown in Table 16.1.

In the chimpanzee *(P. troglodytes)* 18S and 28S cistrons are distributed in five chromosome pairs as in man, but not in the exactly corresponding homologous chromosomes. The difference between these two species relies on two chromosome pairs: one is chromosome 15 that in man carries a NOR, but its homologue in *P. troglodytes* (PTR 16) does not; and neither does it carry satellite stalks or terminal satellites. The other is chromosome 17 in *P. troglodytes* (PTR 17), an acrocentric chromosome with satellite stalks and terminal satellites which carries a NOR; but its human homologue (HSA 18) does not, and neither is it an acrocentric chromosome. In *Pan paniscus* the NORs are distributed in a way similar to those in *P. troglodytes,* but chromosome 23 is metacentric and carries a NOR in its short-arm region as does its acrocentric homologue (No. 23) in *P. troglodytes*. In the gorilla, however, only two pairs of acrocentric chromosomes (Nos. 22 and 23) have a NOR, whereas the large acrocentric chromosomes (Nos. 12, 13, 14, 15, and 16), all of which have satellite stalks and terminal satellites, do not show detectable

NORs. In the orangutan all satellited acrocentric chromosomes carry a NOR at their satellite stalk (9 pairs of chromosomes). It can be observed from Table 16.1 that the distribution of these sequences, although carried by homologous chromosomes between species, is not strictly co-incident when two or more species are compared.

The chromosome distribution of the 18S and 28S rDNA genes in the Hominidae could have been derived either from an ancestral "distribution" of these sequences prior to speciation or from independent amplification in each species. A comparison among species within the Hominidae does not allow us to suggest or to discard any of these two possibilities, one reason why a comparison with other primate species is of interest. In other Hominoid primates such as in the siamang, *Symphalangus syndactylus,* these sequences are located in the short-arm region of a small pair of acrocentric chromosomes (Henderson et al., 1976a). In the white-handed gibbon, *Hylobates lar,* they are also restricted to only one chromosome pair, but in this species all chromosomes are metacentric, and the NOR is located in the secondary constriction region of a medium-sized chromosome (Warburton et al., 1975). This region is variable in size, positively C-banded, and is also involved in association between homologous chromosomes (Tantravahi et al., 1975). A similar distribution was observed in Cercopithecoid monkeys, such as the Rhesus monkey, *Macaca mulatta* (Henderson et al., 1974a); the Colobus monkey, *Colobus pokylomus;* and the baboons, *Papio* spp. *cynocephalus* and *hamadryas* (Henderson et al., 1977). These species, like most Cercopithecoid monkeys, have a distinctive metacentric chromosome pair, or "marker chromosome", in which there is a secondary constriction region frequently associated with nucleoli, which is considered to be the NOR in Cercopithecoid primates (Chiarelli, 1971).

On the other hand, in some platyrrhine monkeys such as the squirrel monkey, *Saimiri sciureus* (Lau and Arrighi, 1976); the owl monkey, *Aotus trivirgatus* (C.K. Miller et al., 1977); and the spider monkey, *Ateles geoffrey* (Henderson et al., 1977), the NOR was also located in a metacentric "marker" chromosome. However, the distribution of 18S and 28S sequences in two other platyrrhines, the saki, *Pithecia pithecia;* and the marmoset, *Sanguinus nigricollis,* was multichromosomal, in three pairs of small metacentric chromosomes and in five pairs of acrocentric chromosomes respectively (Henderson et al., 1977). Finally, in two prosimian species, the tree shrew, *Tupaia tupaia,* and the Lemur, *Lemur fulvis,* the 18S and 28S rDNA genes were detected in two chromosome pairs, and in as many as ten microchromosome pairs, respectively (Henderson et al., 1977). These findings indicated that the amplification of these repetitive sequences in the chromosomes must have been independent in each primate lineage, rather than derived from an ancestral chromosome distribution prior to speciation. This is why the chromosome distribution of the 18S and 28A rDNA genes in the Hominidae is unrelated to the chromosome rearrangements that took place within the group. In some cases, for example in chromosome 16 in the genus *Pan,* we do not find a NOR coinciding with the absence of satellite stalks and terminal satellites in the short arm of this chromosome pair. Although we might then suggest that the NOR has been lost as a consequence of a deletion, it must be mentioned that none of the large acrocentric chromosomes in the gorilla (Nos. 12 to 16), which have satellite stalks and terminal satellites, contains detectable amounts of the 18S and 28S rDNA genes. Similar observations are valid for other rearrangements which are therefore unrelated to the chromosome distribution of these sequences in man and the great apes.

5S rDNA Cistrons in Man and the Great Apes

Contrary to what is found with 18S and 28S sequences, the distribution of 5S sequences in the Hominidae shows a remarkable degree of evolutionary conservation. The 5S rDNA genes in man detected by techniques of in situ hybridisation are located at the terminal region of the long arm of chromosome 1 (Steffensen et al., 1974; Atwood et al., 1975). In the great apes, chromosome 1 has a recognizable homologue in the four species, except that the arm ratio in chromosome 1 of the ape is the opposite to that of man. In chromosome 1 of the ape there is no secondary constriction, thus the long arm of the human chromosome 1 is homologous to the short arm of chromosome 1 in the apes, and the short arm of chromosome 1 in man is homologous to the long arm of chromosome 1 in the apes. The 5S ribosomal cistrons in the great apes are placed at the terminal region of the short arm of chromosome 1, thus at the same site as in man (Henderson et al., 1976b). In the baboon *(Papio cynocephalus),* a species in which a large metacentric chromosome is recognized as homologue to chromosome 1 in man, the location of the 5S rDNA genes was detected at the terminal region of its short arm, a region which is equivalent to the distal region of HSA 1q as demonstrated by chromosome banding (Warburton et al., 1975 in the Baltimore Conference, 1975).

References

Atwood, K.C., Yu, M.T., Johnson, L.D., Henderson, A.S.: The site of 5S RNA genes in human chromosome 1. Cytogenet. Cell Genet. *15*, 50–54 (1975)

Birnstiel, M.L., Grunstein, M.: The ribosomal cistrons of eukaryotes – a model system for the study of evolution of serially repeated genes. FEBS Symp. *23*, 349–365 (1972)

Chiarelli, B.A.: Comparative cytogenetics in primates and its relevance for human cytogenetics. In: Comparative genetics in monkeys, apes and man. Chiarelli, B.A. (ed.), pp. 276–304. London, New York: Academic Press 1971

Croce, C.M., Talavera, A., Basilico, C., Miller, O.J.: Suppression of production of mouse 28S ribosomal RNA in mouse-human hybrids segregating mouse-chromosomes. Proc. Nat. Acad. Sci. USA *74*, 694–697 (1977)

Eliceiri, G.L., Green, H.: Ribosomal RNA synthesis in human-mouse hybrid cells. J. Mol. Biol. *41*, 253–260 (1969)

Evans, H.J., Buckland, R.A., Pardue, M.L.: Location of the genes coding for 18S and 28S ribosomal RNA in the human genome. Chromosoma (Berlin) *48*, 405–426 (1974)

Ferguson-Smith, M.A., Handmaker, S.D.: Observations on the satellited human chromosomes. Lancet *1961 1*, 638–640

Goodpasture, C., Bloom, S.E.: Visualization of nucleolar organizer regions in mammalian chromosomes using silver staining. Chromosoma (Berlin) *53*, 37–50 (1975)

Gosden, J.R., Lawrie, S., Seuánez, H.: Ribosomal and human-homologous repeated DNA distribution in the orangutan *(Pongo pygmaeus)*. Cytogenet. Cell Genet. *21*, 1–10 (1978)

Hatlen, L., Attardi, G.: Proportion of the HeLa cell genome complementary to transfer RNA and 5S RNA. J. Mol. Biol. *56*, 535–553 (1971)

Henderson, A.S., Atwood, K.C., Warburton, D.: Chromosomal distribution of rDNA in *Pan paniscus, Gorilla gorilla beringei* and *Symphalangus syndactylus:* Comparison to related primates. Chromosoma (Berlin) *59*, 147–155 (1976a)

Henderson, A.S., Atwood, K.C., Yu, M.T., Warburton, D.: The site of 5S RNA genes in primates. Chromosoma (Berlin) *56*, 29–32 (1976b)

Henderson, A.S., Warburton, D., Atwood, K.C.: Location of ribosomal DNA in the human chromosome complement. Proc. Nat. Acad. Sci. USA *69*, 3394–3398 (1972)

Henderson, A.S., Warburton, D., Atwood, K.C.: Localization of rDNA in the chromosome complement of the Rhesus *(Macaca mulatta)*. Chromosoma (Berlin) *44*, 367–370 (1974a)

Henderson, A.S., Warburton, D., Atwood, K.C.: Localizing of rDNA in the chimpanzee *(Pan troglodytes)* chromosome complement. Chromosoma (Berlin) *46*, 435–441 (1974b)

Henderson, A.S., Warburton, D., Megraw-Ripley, S., Atwood, K.C.: The chromosomal location of rDNA in selected lower primates. Cytogenet. Cell Genet. *19*, 281–302 (1977)

Hsu, T.C., Spirito, S.E., Pardue, M.L.: Distribution of 18+28S ribosomal genes in mammalian genomes. Chromosoma (Berlin) *53*, 25–36 (1975)

Lau, Y.F., Arrighi, F.E.: Studies of the squirrel monkey, *Saimiri sciureus,* genome. I. Cytological characterizations of chromosomal heterozygosity. Cytogenet. Cell Genet. *17*, 57–60 (1976)

Miller, D.A., Dev, V.G., Tantravahi, R., Miller, O.J.: Suppression of human nucleolus organizer activity in mouse-human somatic hybrid cells. Exp. Cell Res. *101*, 235–243 (1976)

Miller, O.J., Miller, D.A., Dev, V.G., Tantravahi, R., Croce, C.M.: Expression of human and suppression of mouse nucleolar organizer activity in mouse-human somatic cell hybrids. Proc. Nat. Acad. Sci. USA *73*, 4531–4535 (1976)

Miller, C.K., Miller, D.A., Miller, O.J., Tantravahi, R., Reese, R.T.: Banded chromosomes of the owl monkey, *Aotus trivirgatus*. Cytogenet. Cell Genet. *19*, 215–226 (1977)

Steffensen, D.M., Prensky, W., Dufy, P.: Localization of the 5S ribosomal RNA genes in the human genome. In: New Haven Conference (1973). First International Workshop on Human Gene Mapping. Birth Defects: Original Article Series, Vol. X, pp. 153–154. New York: National Foundation 1974

Tantravahi, R., Dev, V.G., Firschein, I.L., Miller, D.A., Miller, O.J.: Karyotype of the gibbons *Hylobates lar* and *H. moloch*. Inversion in chromosome 7. Cytogenet. Cell. Genet. *15*, 92–102 (1975)

Tantravahi, R., Miller, D.A., Dev, V.G., Miller, O.J.: Detection of nucleolar organiser regions in chromosomes of human, chimpanzee, gorilla, orangutan and gibbon. Chromosoma (Berlin) *56*, 15–27 (1976)

Warburton, D., Atwood, K.C., Henderson, A.S.: Variation in the number of genes for rRNA among human acrocentric chromosomes; correlation with frequency of satellite association. Cytogenet. Cell Genet. *17*, 221–230 (1976)

Warburton, D., Henderson, A.S., Atwood, K.C.: Localization of rDNA and giemsa banded chromosome complement of white-handed gibbon, *Hylobates lar*. Chromosoma (Berlin) *51*, 35–40 (1975)

Warburton, D., Yu, M.T., Atwood, K.C., Henderson, A.S.: The location of RN5S genes in the chromosomes of primates. Baltimore Conference (1975): Third International Workshop on Human Gene Mapping. Birth Defects: Original Article Series XII, 7. New York: National Foundation 1976

Chapter 17 Late DNA Replicating Patterns in the Chromosomes of Man and the Great Apes

DNA Replication at the Chromosome Level

Studies of the pattern of DNA replication have been a useful approach for analysing the timing of DNA synthesis in relation to the cell cycle. The method of labelling chromosomes with a radioactive precursor (tritiated thymidine) was first introduced by Taylor et al. (1957), and has been used to study the chromosome complement of many organisms, including man (for a review see Passarge, 1970). One of the most important findings before the development of banding techniques was that the asynchrony of DNA replication within a chromosome group permitted the presumptive identification of a pair of homologues, based on a specific pattern of labelling. It was also evident in man that the Y chromosome in males and one of the X chromosomes in females were late-replicating in relation to the rest of the chromosome complement. Using similar methods, Low and Benirschke (1969) studied the late-replication pattern of the chromosome complement of *Pan troglodytes* and compared it with that of man. Although many chromosomes of man and chimpanzee showed a very similar morphology, Low and Benirschke (1969) found considerable differences in labelling pattern.

DNA Replication Sites in Relation to Chromosome Banding

It was not until the development of chromosome banding techniques that a more detailed comparison of these regions became possible. Ganner and Evans (1971) and Calderon and Schnedl (1973) studied the patterns of DNA replication in relation to the banding patterns of the human chromosome complement. These studies showed that regions of late DNA replication coincided with regions of positive G- and C-banding, as well as with regions of high fluorescence intensity.

The patterns of DNA replication were better understood when BUdr (5-bromodeoxyuridine) was used as a thymidine analogue. Although BUdr had been used in the past (Palmer and Funderburk, 1965), its effect on chromosome structure was not fully understood until Zakharov and Egolina (1972) demonstrated that chromosome regions incorporating BUdr exhibited increased despiralization. This effect was dose-dependent and inversely proportional to the time between BUdr incorporation and the onset of mitosis. In 1973 Latt reported that BUdr had the property of quenching fluorescence when the chromosomes were stained with Hoechst 33258, a fluorescent dye. But he noticed that the incorporation of thymidine, 5 h before harvesting a cell culture that had undergone one round of replication with BUdr, had the effect of enhancing fluorescence with Hoechst 33258, in the regions where thymidine was incorporated. This was demonstrated using tritiated thymidine and autoradiographing the same cells after Hoechst staining. Thus, the differential staining produced a banding pattern, the regions of brighter fluorescence being also those of late replication. These regions coincided in general with regions of positive G-banding of the human chromosome complement; one exception, however, was the secondary constriction of chromosome 9, which is negatively G-banded. But not all positively G-banded areas were late-replicating. The reverse procedure, i.e., the late incorporation of BUdr 5 h before harvesting a cell culture grown in a medium with thymidine, produced a reverse, R-like band pattern; late-replicating regions showed quenched fluorescence, while early-replicating regions appeared brighter (Latt, 1973). The procedure described by Latt (1973) was later modified by Perry and Wolf (1974) using Giemsa staining and $2 \times$ SSC denaturation which allows permanent preparations to be made. Chromosome regions incorporating BUdr appear faintly stained. This method permitted a more precise analysis of the chromosome structure, especially in relation to the occurrence of sister chromatid exchanges (SCE) in cells undergoing more than one round of replication with BUdr. The technique of Perry and Wolf (1974) was used by Grzeschik et al. (1975), Epplen et al. (1975), and Kim et al. (1975) to demonstrate the late-replicating pattern of the human chromosome complement in individual chromosome bands, thus with greater precision than by employing autoradiographic techniques. These studies confirmed the findings of Latt (1973), that regions of late replication corresponded generally to regions of positive G-banding (Fig. 17.1), while early-replicating regions coincided with R-band regions.

In the great apes, chromosomes that had incorporated thymidine for the final 5 h before harvesting (T-pulsed cultures), after one round of replication with BUdr also produced a pattern in which late-replicating regions coincided with Q-, G-, and C-bands as in man (Figs. 17.1 and 17.2). BUdr (B) pulsed cultures, on the other hand, showed a reverse

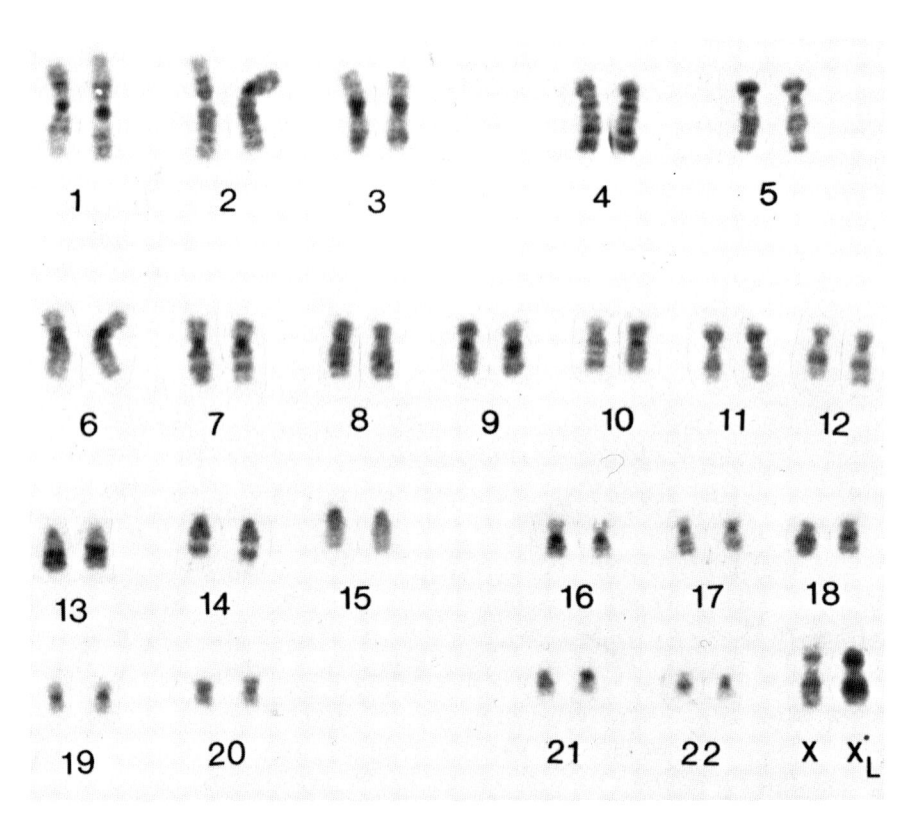

Fig. 17.1. Late-replicating pattern of human chromosomes in a female individual. Cultures were grown in culture medium containing BUdr for 43 h, and were pulsed with thymidine for the 5 following hours and harvested at 48 h. *Darkly stained regions* correspond to late-replicating regions and coincide with regions of positive Q-, G-, and C-banding. X_L denotes the late replicating X chromosome

pattern; late-replicating regions coincided with R-bands (Figs. 17.3 and 17.4; Seuánez, 1977). When species were compared, it was found that the Q-, G-, or C-band regions, late-replicating in one species, were the same as those late-replicating in other species. A good example of this is found in chromosome 1 of man and its ape homologues. In HSA 1 there is region which is late-replicating, but which is absent in the ape homologues (the secondary constriction). However, other regions of the human chromosome 1 do have homologies in the apes chromosome 1, i.e., regions p 35, p 33, p 31, p 21, q 31, q 41, and q 43, and some of these can be seen to be late-replicating in all species investigated (Figs. 17.1, 17.2, 17.3, and 17.4). Findings of this kind are best observed in chromosomes where a good homology of Q- and G-banding exists between different species, and where a chromosome band of a human chromosome

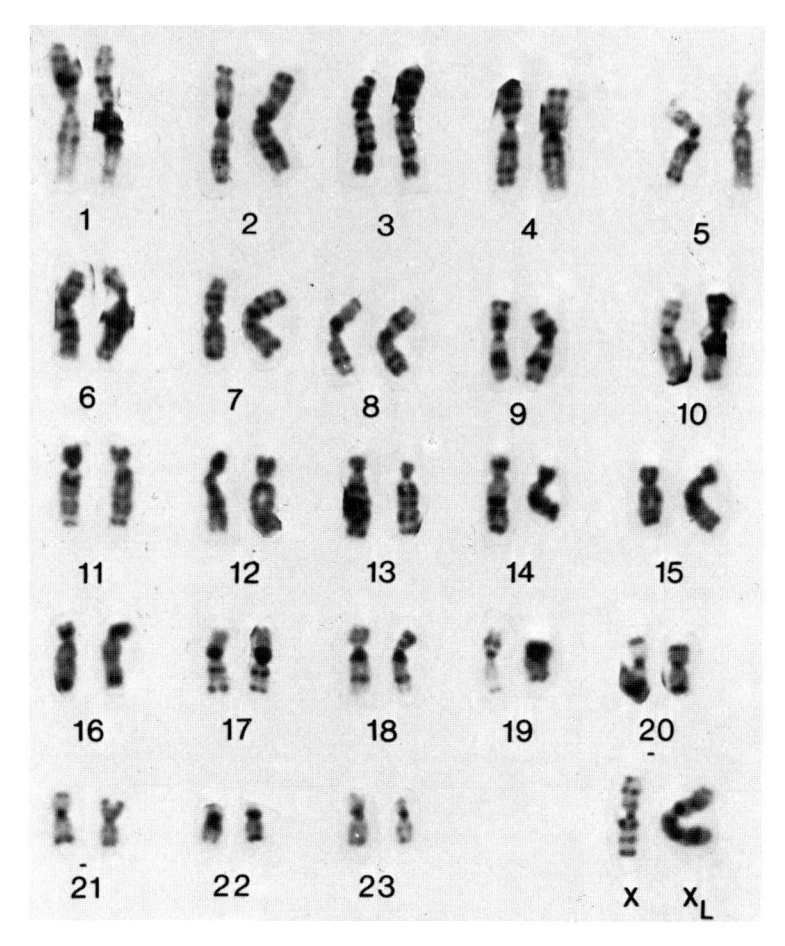

Fig. 17.2. Late-replicating pattern of gorilla chromosomes in a female individual. Cultures were grown as indicated in Figure 17.1, and late-replicating regions coincide with regions of positive Q-, G-, and C-banding. X_L denotes late-replicating X chromosome

can be precisely found in an ape chromosome. An example of this remarkable similarity is chromosome 13 in man and its ape homologues (PTR 14, GGO 14, and PPY 14). In man there are two clear regions of late DNA replication in the long arm of chromosome 13, that correspond to bands q 21 and q 31. These two regions are separated by a region of early DNA replication corresponding to q 22, a region of negative G-banding (Fig. 17.5). In PTR 14, GGO 14, and PPY 14 also the same two regions are late-replicating. This was clearly confirmed using cell cultures from a specimen of *Gorilla gorilla* in which a marker chromosome 14 was found [del(14)p 31→p 11:], so that this chromosome could be positively distinguished by its morphology from any other acrocentric chromosome

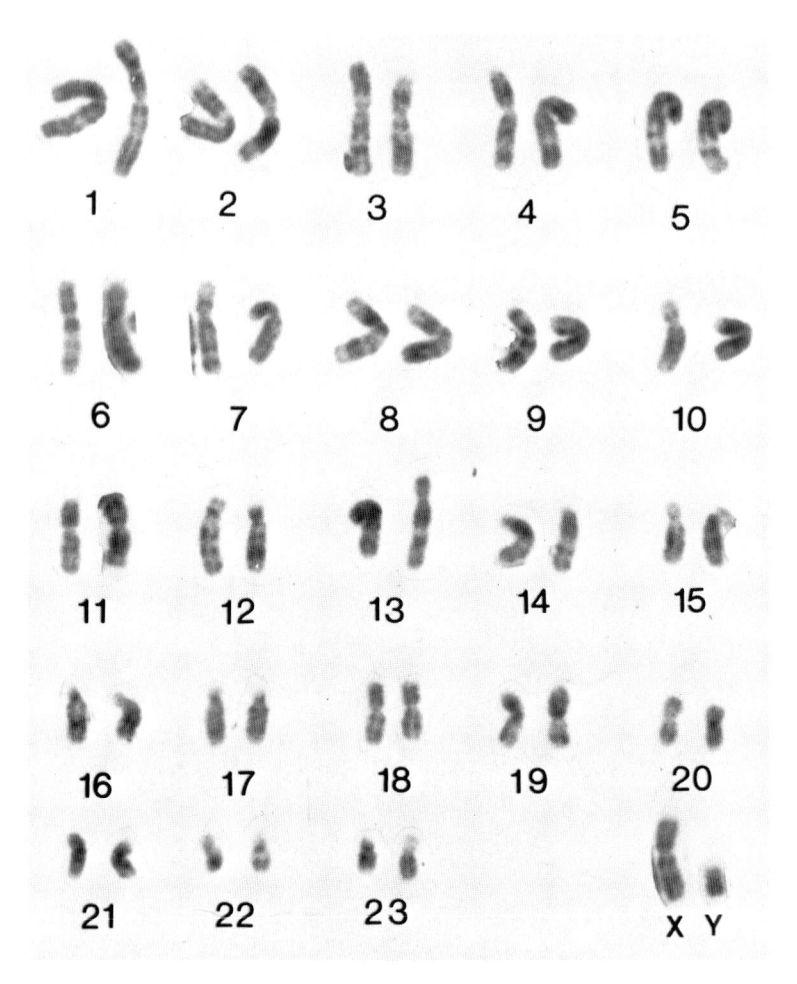

Fig. 17.3. Chromosomes of the chimpanzee (spp. *troglodytes*) in a male individual. Cultures were grown in medium containing thymidine for 67 h, pulsed with BUdr for the 5 following hours, and harvested at 72 h. Late-replicating regions appear pale; they coincide with R-band regions

(Seuánez, 1977). All brilliant fluorescent regions in *Pan troglodytes* and *Gorilla gorilla* correspond to regions of late DNA replication. In *Pongo pygmaeus*, the chromosome regions homologous to those showing brilliant fluorescence in other species were also late-replicating (e.g., the short arm of chromosome 14 in *Pongo pygmaeus*, which is homologous to the short-arm region of chromosome 13 in man and 14 in *Pan troglodytes* and *Gorilla gorilla*).

Late-replicating regions of the chromosome complement of man and these three species of great apes also coincided with regions of positive

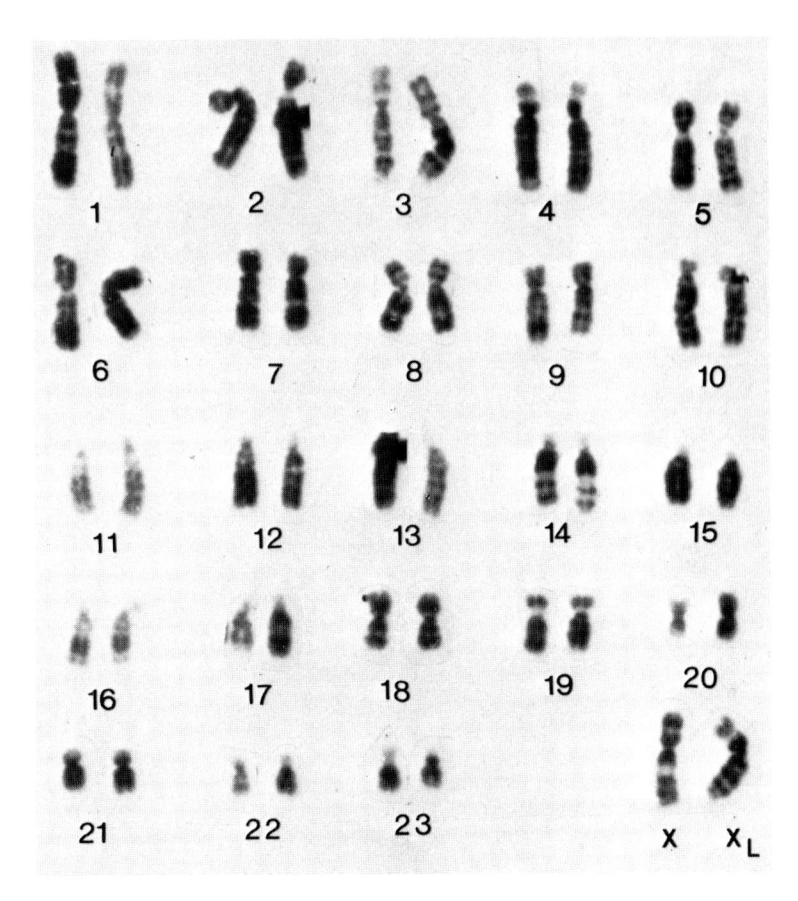

Fig. 17.4. Chromosomes of the orangutan in a female individual. Cultures were grown as indicated in Figure 17.3. Late-replicating regions appear pale; they coincide with R-band regions. X_L denotes the late-replicating X chromosome

C-banding. This was the case with the secondary constriction of chromosome 1, 9, and 16 in man, and 17 and 18 in *Gorilla gorilla* (Figs. 17.1 and 17.2). Another example was that of the Y chromosome heterochromatic regions of all species, and the telomeric regions of many chromosome arms in *Pan troglodytes* and *Gorilla gorilla*. It was common to find situations in which one terminal region was late-replicating in one homologue, whereas in the other homologue of the same individual this was not clearly evident. This was interpreted as actual differences between homologues (kinetic polymorphisms) because a comparison with the C-band karyotype of the same animal showed that the terminal heterochromatic regions were present in the two homologous chromosomes. The absence of the late-replicating terminal region in one

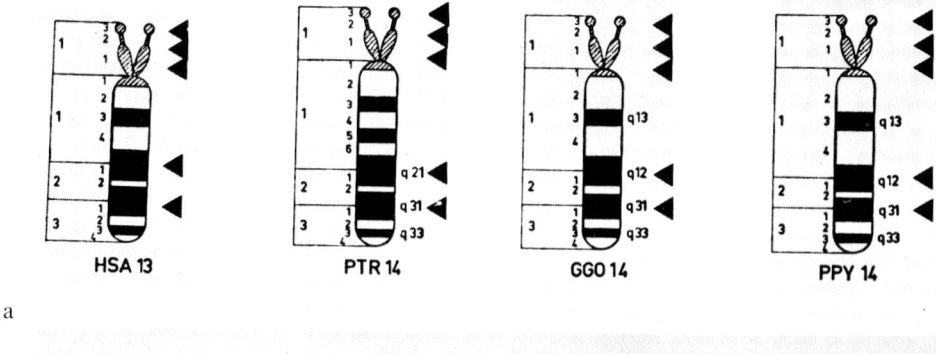

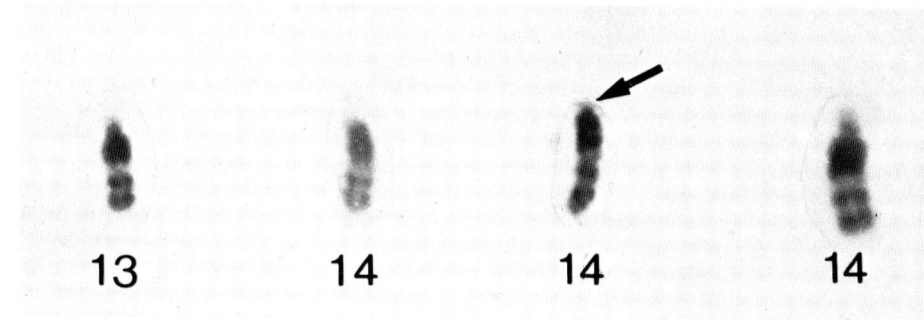

Fig. 17.5. a G-band diagrams of chromosome 13 in man and its ape homologues as shown in the Paris Conference, 1971; supplement 1975. *Arrows* point to late replicating regions which are shown in **b**, from cells grown in medium containing thymidine and pulsed with BUdr before harvesting. Late-replicating regions appear pale. Note the remarkable similarity in late-replicating patterns between species. *Arrow* points to marker chromosome

homologue could not therefore be due to the absence of this region from that chromosome.

The X Chromosome

Differences in the late-replicating pattern of the X chromosomes were observed in female animals of these species, identical to those found in human female cells. In T-pulsed cells one of the X chromosomes appeared more intensely stained than the other, and the regions of intense staining corresponded to regions of positive G-banding, probably bands p21, q21, and q23 (Figs. 17.1 to 17.4). The extreme tip of the long arm (band q27) of this chromosome was generally pale. The resolution at the band level was, however, poor in T-pulsed cells, since the intensity of staining was very high. The other X chromosome showed the same

Fig. 17.6. Late-replicating patterns of the normal-replicating, (X), and the late-replicating X chromosome, (X_L), in female individuals in man (*HSA*), chimpanzee (*PTR*), gorilla (*GGO*), and orangutan (*PPY*). Cells were grown in medium containing thymidine and pulsed with BUdr before harvesting. Late-replicating regions appear pale

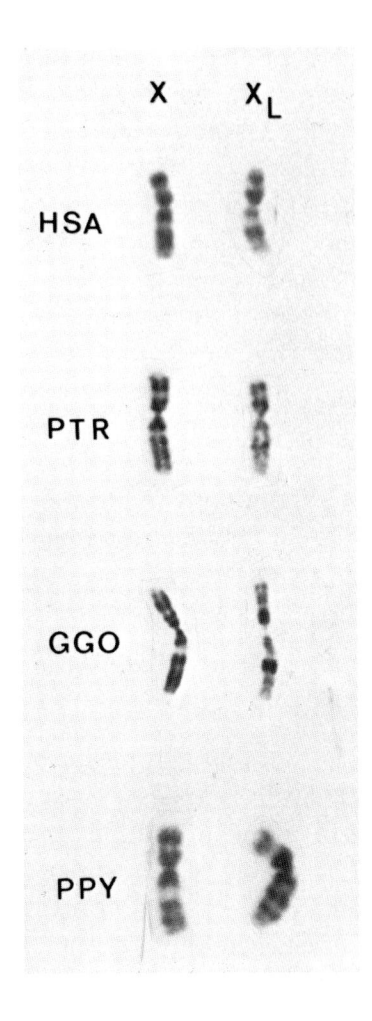

pattern of late replication as the X chromosome in male cells, with a region of late replication at the short arm (p 21) and another at the long arm (q 21). The size of these two regions was clearly smaller than that of the more intensely stained X chromosome. From previous experiments in which cells had been grown in BUdr, but pulse-labelled with tritiated thymidine, we know that the X chromosome showing intense Giemsa staining corresponds to the late-replicating X chromosome, whereas the other corresponds to the normally replicating X chromosome (Latt, 1974). A better understanding of the sites of DNA replication in the X chromosomes was obtained, however, from B-pulsed cultures. In B-pulsed cells, one X chromosome appeared very pale compared with the rest of the chromosome complement (Fig. 17.6), whereas the other X chromosome showed an identical replication pattern to that observed in the X chromosome of B-labelled male cells. The very pale X chromosome was obviously the late-replicating X chromosome, as was demonstrated by Latt (1974) in cell cultures pulsed for the final 5 h with tritiated BUdr. A close examination of the sequence of DNA replication in this chromosome shows that it was pale (late-replicating), except for two regions: one at the proximal segment of the short arm and another at the middle of the long arm. These regions coincided with regions of negative G-banding of the X chromosome (p 11 and q 22). It must be noticed that the terminal region of the long arm was also pale (late-replicating). This was contrary to expectation, since the terminal region of the long arm of the late-replicating X chromosome in most of the T-pulsed cells was pale (early-replicating), a result for which we found no explanation.

Euchromatin, Heterochromatin and DNA Replication

The fact that late-replicating regions coincided with G-, Q-, and C-band regions is of interest since we know that in man heterochromatin is late-replicating in relation to euchromatin (Lima de Faria and Jaworska, 1968). Thus, the observation that regions of positive Q- and G-banding are late-replicating in man and the great apes may suggest that Q- and G-bands in these species are also heterochromatin. The definition of heterochromatin, as initially stated by Heitz (1928), applied to those regions in the chromosome complement of organisms showing a preferential staining property, and remaining condensed throughout the entire cell cycle. More recently, however, the definition of heterochromatin has been widened to include two kinds of heterochromatin (Brown, 1966). One called "facultative" refers to the behaviour of one X chromosome which has become heterochromatic and genetically inactive in the somatic cells of mammalian species at some stage of embryonic development (Lyon, 1961), while remaining euchromatic in the germ cell line. A second, "constitutive", heterochromatin is that which according to Arrighi et al. (1974) has the following properties:

1) It remains condensed throughout the entire cell cycle except presumably during replication
2) It can be demonstrated by C-banding

3) It shows a definite and consistent pattern of distribution within the karyotype
4) It is genetically inert
5) It is usually late-replicating in S-phase
6) It contains a large amount of repetitive DNA sequences (including satellite DNA).

Still, it must be stressed that this definition of "constitutive" heterochromatin is imprecise when extended to all organisms. In man, for example, the cytological demonstration of constitutive heterochromatin is evident with C-banding; but some regions of positive C-banding can also be demonstrated with G-banding, as the secondary constrictions of chromosome 1 and 16, especially when G-banding is obtained with proteolytic enzymatic digestion. In other regions of the human chromosome complement G- and C-banding are mutually exclusive, as in the secondary constriction of chromosome 9 which is negatively G-banded and positively C-banded. In *Pan troglodytes* the interstitial C-band regions in chromosome 6 and 14 also correspond to positive Q- and G-band regions, whereas the terminal C-band regions of *Pan troglodytes* and *Gorilla gorilla* are negatively G-banded, but positively Q-banded. Thus, the strict differentiation between C- and G- (and sometimes Q-) band regions is not possible, and it might well be that G-band regions represent a different stage of heterochromatin than C-band regions, as has been proposed by Comings (1974).

The relation between chromosome banding and time of DNA replication is also conflicting. Firstly, similar chromosome-banding patterns are obtained from cells of different tissues in man (e.g., lymphocytes and fibroblasts), although the pattern of late DNA replication in these two types of human cells has been reported to be different (Slezinger and Prokofieva-Belgovskaya, 1968). A similar situation was obtained also between lymphocytes and amniotic cells in culture (German and Aronian, 1971). Secondly, while in human lymphocytes some late replicating regions coincide with G- and C-bands (Fig. 17.1), in other organisms such as the Seba's fruit bat, *Carollia perspicillata* (Pathak et al., 1973), and in two species of kangaroo rat, *Dipodomys merriani* and *Dipodomys panamintinus* (Bostock and Christie, 1974, 1975) G- and Q-band regions have been found to start and complete their replication before other non-Q- and non-G-band regions. However, it has also been reported that, in the same species of kangaroo rat (Bostock and Christie, 1974, 1975), in the mouse, *Mus musculus* (Hsu and Arrighi, 1971; Hsu and

Markvong, 1975), and in the Indian muntjack, *Muntiacus muntjack* (Sharma and Dhaliwal, 1974), some C-band regions start replicating earlier than other non-C-band regions. Moreover, in the Chinese hamster, *Cricetulus griseus,* it has been shown that G- and non-G-band regions may replicate at different stages of the S-phase (Stubblefield, 1975). Thus, a straightforward relationship between chromosome bands and late DNA-replicating sites is unlikely. Besides, there is no straightforward coincidence between late-replicating regions and sites where satellite DNA sequences have been localized in man (Chap. 15), and the same applies to other species.

In man, for example, many late-replicating regions do not contain detectable amounts of satellite DNA sequences (e.g., bands q 21 and q 23 in chromosome 13). In *Pan troglodytes,* for example, the proximal region of the long arm of chromosome 6 was not late replicating (Fig. 17.3), and the pattern of late replication in this chromosome was identical to that in chromosome 7 of man where no major amounts of satellite DNA sequences had been detected (Chap. 15). In the Chinese hamster, the long arm of the X chromosome and the entire Y chromosome are heterochromatic and late-replicating (Taylor, 1960; Stubblefield, 1975), but do not contain satellite DNA (Arrighi et al., 1974). Furthermore, a comparison with other organisms shows that satellite DNA is not necessarily the last replicating fraction of the mammalian genome. In the mouse, for example, Tobia et al. (1970) showed that 80% of the satellite DNA was synthesized after main band DNA, but there is also evidence that mouse satellite DNA terminates synthesis in the third quarter of S-phase (Bostock and Prescott, 1971). In one species of kangaroo rat *(D. merriani)* it was shown that one satellite fraction (MS) was early-replicating as was G-C-rich main band DNA (Bostock and Christie, 1974). It can then be concluded that, although replication of mammalian DNA in chromosomes is a co-ordinated process involving sequential replication of many heterogeneous DNA molecules, it has no straightforward explanation in terms of the physical and biological properties of DNA molecules, or in terms of chromosome regions demonstrated by chromosome banding.

So the similar pattern of late DNA replication observed in the chromosome complement of man and the great apes must have resulted from the conservation of the mechanisms of replication during phylogeny. There are few reports published so far in which the late-replicating patterns of the chromosomes of two or more phylogenetically related mammalian species showing closely similar karyotypes have been compared. The species of kangaroo rat studied by Bostock and Christie (1974, 1975) show remarkable differences in their karyotypes, and it is difficult to infer interspecific chromosome homologies, as we may do between man and the great apes.

Hsu and Markvong (1975) studied the late-replicating pattern of three species and one subspecies of mouse (*Mus musculus, Mus fulvidiventris, Mus dunni,* and *Mus musculus mollosinus*). These have a diploid chromosome number of 40, with similar Q- and G-banding patterns; differences between species and subspecies being due to the size of the heterochromatic regions and of the sex chromosomes (Markvong et al., 1975). The late-replicating patterns of *Mus musculus* and *Mus m. mollosinus* were identical; their centromeric regions were not the last to replicate in S-phase, and the Y chromosome finished replication before some autosomal regions. In *Mus fulvidiventris,* however, the centromeric regions were the last to complete replication, but its smaller Y chromosome replicated earlier than the larger Y chromosome of *Mus musculus*. In *Mus dunni* the heterochromatic short arms of the autosomes were not late-replicating, but the heterochromatic region of the X chromosome and the very long Y chromosome were. This indicated that not only the amount of constitutive heterochromatin has changed in the genus *Mus,* but also its sequential DNA replication, although the diploid chromosome number and the Q- and G-banding patterns have been basically conserved. This finding apparently contrasts with ours in the Hominidae, in which a constant pattern of sequential replication is observed between different species, although a greater variability is observed in chromosome number and morphology.

It would not be surprising if the difference observed within the genus *Mus* is a consequence of the rapid evolutionary divergence of these species as this divergence usually occurs in the Rodentia, in which reproductive performance is high, sexual maturity occurs early in life, litters are large, and animals live within restricted environments (see Chap. 8). And Hsu and Markvong (1975) have suggested that these changes in replication patterns might result from a divergence in satellite DNA sequences in the heterochromatic regions of the genus *Mus*.

The factors that control the sequential replication of chromosomes are unknown, and there is a standing controversy whether the sequential DNA replication of each chromosome is directly self-controlled (by carrying the genetic information for initiation molecules which would act on its own replicons), or is mediated by the cytoplasm. Evidence in favour of cytoplasmic control is given by the fact that one X chromosome in all female mammals becomes inactivated at some stage of embryonic development (Lyon, 1961), and that late-replicating patterns of chromosomes of cells from different tissues vary in man (Slezinger and Prokofieva-Belgovskaya, 1968; German and Aronian, 1971). However, there is also evidence that each chromosome may control its own sequential replication in man, since in human–mouse hybrid cells in which many human chromosomes were randomly lost, the sequential pattern of replication of the remaining human chromosomes was unchanged from that of the parental cell line. This suggested that the conservation of the pattern of sequential DNA replication did not require an intact human genome, and besides, it was apparently independent of which chromosome had been eliminated (Lin and Davison, 1975). Graves (1975) has observed that hybrid cells of two species with different S-phase duration start S-phase synchronously, but each chromosome set has a similar S-phase period to the one it had in the parental cell line. A common pattern of chromosome replication between phylogenetically related species would result, if the chromosomes of these species kept the genetic information for controlling their own sequential replication as in the common ancestor of the group. Thus, such conservation would be in a way comparable to the conservation of the original structural gene loci in the Hominidae discussed in Chap. 12.

References

Arrighi, F. E., Hsu, T. C., Pathak, S., Sawada, H.: The sex chromosomes of the Chinese hamster: constitutive heterochromatin deficient in repetitive DNA sequences. Cytogenet. Cell Genet. *13*, 268–274 (1974)

Bostock, C. J., Christie, S.: Chromosome banding and DNA replication studies on a cell line of *Dipodomys merriani*. Chromosoma (Berlin) *48*, 73–87 (1974)

Bostock, C. J., Christie, S.: Chromosomes of a cell line of *Dipodomys panamintinus* (Kangaroo rat). A banding and autoradiographic study. Chromosoma (Berlin) *51*, 25–34 (1975)

Bostock, C. J., Prescott, D. M.: Buoyant density of DNA synthesised at different stages of the S phase of mouse α-cells. Exp. Cell Res. *64*, 267–274 (1971)

Brown, S. W.: Heterochromatin. Science *151*, 417–425 (1966)

Calderón, D., Schnedl, W.: A comparison between quinacrine fluorescence banding and ³H-thymidine incorporation. Humangenetik *18*, 63–70 (1973)

Comings, D. E.: Structure of mammalian chromosomes. In: Physiology and genetics of reproduction (part A). Coutinho, E., Fuchs, F. (eds.), pp. 19–27. New York, London: Plenum Press 1974

Epplen, J. T., Siebers, J. N., Vogel, W.: DNA replication patterns of human chromosomes from fibroblasts and amniotic fluid cells revealed by a giemsa staining technique. Cytogenet. Cell Genet. *15*, 177–185 (1975)

Ganner, E., Evans, H. J.: The relationship between patterns of DNA replication and of quinacrine fluorescence in the human chromosome complement. Chromosoma (Berlin) *35*, 326–341 (1971)

German, J., Aronian, D.: Autoradiographic studies of human chromosomes. VI. Comparison of the end-of-S-patterns in lymphocites and amniotic epithelial cells. Chromosoma (Berlin) 35, 99–110 (1971)

Graves, J.M.: Controls of DNA synthesis in somatic cell hybrids. In: The eukaryote chromosome. Peacock, W.J., Brock, R.D. (eds.), pp. 367–379. Canberra: Australia National Univ. Press 1975

Grzeschik, R.H., Kim, M.A., Johannsmann, R.: Late replicating bands in human chromosomes demonstrated by fluorochrome and giemsa banding. Humangenetik 29, 41–59 (1975)

Heitz, E.: Das Heterochromatin der Moose. Jahrbuch für wissenschaftliche Botanik 69, 762–818 (1928)

Hsu, T.C., Arrighi, F.E.: Distribution of constitutive heterochromatin in mammalian chromosomes. Chromosoma (Berlin) 34, 243–253 (1971)

Hsu, T.C., Markvong, A.: Chromosomes and DNA of *Mus:* Terminal DNA synthetic sequences in three species. Chromosoma (Berlin) 51, 311–322 (1975)

Kim, M.A., Johannsmann, R., Grzeschick, R.H.: Giemsa staining of the sites replicating DNA early in human lymphocyte chromosomes. Cytogenet. Cell Genet. 15, 363–371 (1975)

Latt, S.: Microfluorometric detection of deoxyribonucleic acid replication in human metaphase chromosomes. Proc. Nat. Acad. Sci. USA 70, 3395–3399 (1973)

Latt, S.: Microfluorometric analysis of DNA replication in human X chromosomes. Exp. Cell Res. 86, 412–415 (1974)

Lima de Faria, A., Jaworska, H.: Late DNA synthesis in heterochromatin. Nature (London) 217, 138–142 (1968)

Lin, M.S., Davison, R.L.: Replication fo human chromosomes in human-mouse hybrids: Evidence that the timing of DNA synthesis is determined independently in each human chromosome. Somat. Cell Genet. 1, 111–122 (1975)

Low, R.J., Benirschke, K.: The replicating pattern of the chromosomes of *Pan troglodytes*. 2nd Congr. Primatol. Vol. 2, pp. 95–102. Basel, New York: Karger 1969

Lyon, M.F.: Gene action in the X chromosomes of the mouse *(Mus musculus)* Nature (London) 190, 372–373 (1961)

Markvong, A., Marshall, J.T., Pathak, S., Hsu, T.C.: Chromosomes and DNA of *Mus:* the karyotypes of *M. fulvidiventris* and *M. dunni*. Cytogenet. Cell Genet. 14, 116–125 (1975)

Palmer, C.G., Funderburk, S.: Secondary constrictions in human chromosomes. Cytogenetics 4, 261–276 (1965)

Passarge, E.: Der Karyotyp des Menschen. In: Methoden in der medizinischen Cytogenetik. Schwarzacher, H.G., Wolf, U. (eds.), pp. 86–152. Berlin, Heidelberg, New York: Springer 1970

Pathak, S., Hsu, T.C., Utakoji, T.: Relationships between patterns of chromosome banding and DNA synthetic sequences: a study on the chromosomes of the Seba's fruit bat, *Carollia perspicillata*. Cytogenet. Cell Genet. 12, 157–164 (1973)

Perry, P., Wolf, S.: New giemsa method for the differential staining of sister chromatids. Nature (London) 251, 156–158 (1974)

Seuánez, H.N.: Chromosomes and spermatozoa of the great apes and man. Thesis, Univ. Edinburgh (1977)

Sharma, T., Dhaliwal, M.K.: Relationship between patterns of late S DNA synthesis and C- and G-banding muntjak chromosomes. Exp. Cell Res. 87, 394–397 (1974)

Slezinger, S.I., Prokofieva-Belgovskaya, A.: Replication of human chromosomes in primary cultures of embryonic fibroblasts. I. Interchromosomal asynchrony of DNA replication. Cytogenetics 7, 337–346 (1968)

Stubblefield, E.: Analysis of the replication pattern of Chinese hamster chromosomes using 5-Bromodeoxyuridine suppression of 33258 Hoechst Fluorescence. Chromosoma (Berlin) 53, 209–221 (1975)

Taylor, J.H.: Asynchronous duplication of chromosomes in cultured cells of Chinese hamster. J. Biophys. Biochem. Cytol. 7, 455–464 (1960)

Taylor, J.H., Woods, P.S., Hughes, W.L.: The organization and duplication of chromosomes as revealed by autoradiographic studies using tritium labelled thymidine. Proc. Nat. Acad. Sci. USA *43*, 122–128 (1957)

Tobia, A., Schildkraut, C.L., Maio, J.J.: DNA replication in synchronised cultured mammalian cells. I. Time of synthesis of molecules of different average guanine and cytosine content. J. Mol. Biol. *54*, 499–515 (1970)

Zakharov, A.F., Egolina, N.A.: Differential spiralization along mammalian mitotic chromosomes. I. BUdr-revealed differentiation in Chinese hamster chromosomes. Chromosoma (Berlin) *38*, 341–365 (1972)

Chapter 18 Evolution of Genome Size in Man and the Great Apes

The DNA Content of Man and Other Organisms

Evolution of eukaryotes from prokaryotes has obviously been accomplished with a substantial increase in genome size. The bacterial genome, for example, accounts for approximately 0.007 pg of DNA (1 pg = 10^{-12} g), while haploid cells of salamanders and some plants may contain up to 100 pg (Sparrow et al., 1972). DNA measurements in organisms of the same species have demonstrated that DNA content is constant among individuals that are chromosomally normal (Vendrely and Vendrely, 1948). Comparisons among individuals of different species have shown that phylogenetically related species have similar DNA content (Mirsky and Ris, 1951). In man, for example, the DNA content of diploid cells has been estimated to be 6 pg (Vendrely and Vendrely, 1957), and a comparison with other species has shown that the DNA content of human diploid cells is similar to that of placental mammals (Atkin et al., 1965).

If man is compared to other primates, among them the great apes, some interesting findings are evident. When Manfredi-Romanini (1972, 1973), using microdensitometric techniques on Feulgen-stained cells, compared the DNA content of human lymphocytes with those of 33 other non-human primate species, she found that man, the great apes, the Cercopithecoidea and the Ceboidea have conserved, within close limits, similar amounts of DNA. The Hylobatidae, however, appeared to be an exception, in showing the lowest DNA content of all the primates tested, and three Cercopithecoid species, *Cercocebus torquatus, C. galeritus* and *Nasalis nervatus*, showed higher values than those found in seven other species of this group. A comparison between man and three species of great ape showed that man had the lowest DNA content of the Hominidae; the direction in which DNA was found to increase was man < gorilla < chimpanzee < orangutan.

More recently, comparative studies of the DNA content of human and great ape spermatozoa have been reported (Seuánez et al., 1977), using techniques for estimating total dry mass (TDM) of the spermatozoal head with an integrating microinterferometer. The choice of mature spermatozoa as the kind of cell used in measurements has the advantage that their DNA content is stable and genetically inactive, a situation which does not exist in most somatic cells. DNA synthesis in the meiotic cycle of both plants and animals occurs at the S-phase – prior to the meiotic prophase (Taylor and McMaster, 1954). Although a small additional DNA synthesis occurs during the prophase of the first meiotic division in plants (Hotta et al., 1966), the mouse (Hotta et al., 1977), and also in man (Lima de Faria et al., 1968), all the DNA synthesis is completed before metaphase of the first meiotic division. Thus, the amount of DNA which is contained in the spermatids and spermatozoa will depend only on how chromosomes segregate and divide during meiosis. Since the amount of DNA does not increase or decrease during spermiogenesis (the maturation of spermatids into spermatozoa) the DNA content of spermatids and spermatozoa remains constant and does not fluctuate (Monesi, 1971). Another advantage of using interferometric techniques to determine the TDM of the sperm head is that measurements are not affected by the degree of chromatin condensation, as are DNA measurements obtained from Feulgen stained cells with microdensitometry (Bedi and Goldstein, 1974). It must be pointed out that, since

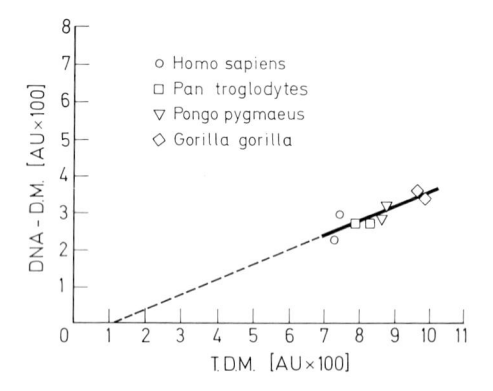

Fig. 18.1. To test whether TDM is proportional to DNA content, 10 spermatozoa from each of two individuals from each species (except *Pan paniscus*) were measured for TDM. After treatment with 5% trichloracetic acid at 90 °C to remove DNA, the same spermatozoa were relocated and measured for dry mass after extraction (DMAE). Slides were then stained with Feulgen and the same spermatozoa measured for residual DNA with the M-86 instrument in the microdensitometric mode. Only if no DNA was detected were the DMAE values considered valid. The minimum extraction time needed to remove all DNA was 15 min in man and 30 min in the great apes. The difference between TDM and DMAE represents the dry mass of extracted DNA (DNA-DM). The figure shows mean values of DNA-DM and TDM for each individual, and the estimated least-squares regression line. The latter passes close to the origin, suggesting that proportional changes in TDM are identical to those for DNA-DM. (The line should not of course be interpreted as a valid extrapolation beyond the range of the plotted points.) ○ *Homo sapiens*; □ *Pan troglodytes*; ▽ *Pongo pygmaeus*; ◇ *Gorilla gorilla*. [Seuánez et al., Nature (London) 270, 343–345 (1977)]

measurements of TDM represent the total amount of substance present in the spermatozoal head (protoplasm), they are not specific for any substance such as DNA. Nonetheless, values of TDM have been found to be proportional to the DNA content of the spermatozoal head in man, *Pan troglodytes, Gorilla gorilla* and *Pongo pygmaeus* (Fig. 18.1) a fact that allows direct comparisons to be made between dry mass measurements of different species as indicative of their genome size (Seuánez et al., 1977). These measurements show a distribution around a mean value for each of the species tested (Fig. 18.2): the mean values of all species being significantly different from each other, except those of *Pan troglodytes* and *Pan paniscus*. Such observed differences between mean TDM values were not due to technical artifacts and, moreover, could not be explained by deviations from expected mean values due to aneuploidy (Table 18.1). Thus, they reflected actual differences in genome size among species.

Why Has DNA Content Changed?

Experiments showed (Seuánez et al., 1977) that the genome size of man, estimated by the mean value of haploid spermatozoa, was significantly lower than that of any other species in the Hominidae, and this largely coincided with the findings of Manfredi-Romanini (1972, 1973). But, the species sequence in which DNA content increased was different, being man $<$ *Pan troglodytes* $<$ *Pongo pygmaeus* $<$ *Gorilla gorilla*. Why man shows less DNA in his genome than the great apes is not yet clearly explained, but it is important to consider some of the evolutionary implications of genome size variation observed in other species. Firstly, it has been found that DNA content among species belonging to the same taxon (e.g., a family or a class) may show differences which are somehow related to their degree of structural complexity or specialization (Hinegardner, 1976). For example, the more generalized species (those showing more characteristics in common with other members of the taxon) usually contain more DNA than those that are more specialized (i.e., those that have characteristics which are rare among other members of the same taxon). The rule seems to be then, that specialization is accompanied by a decrease in DNA, since each step of specialization is achieved by selection operating on DNA sequences already existing in the genome. When successful adaptations occur and organisms become more specialized, other DNA sequences are released from selective pressures, and

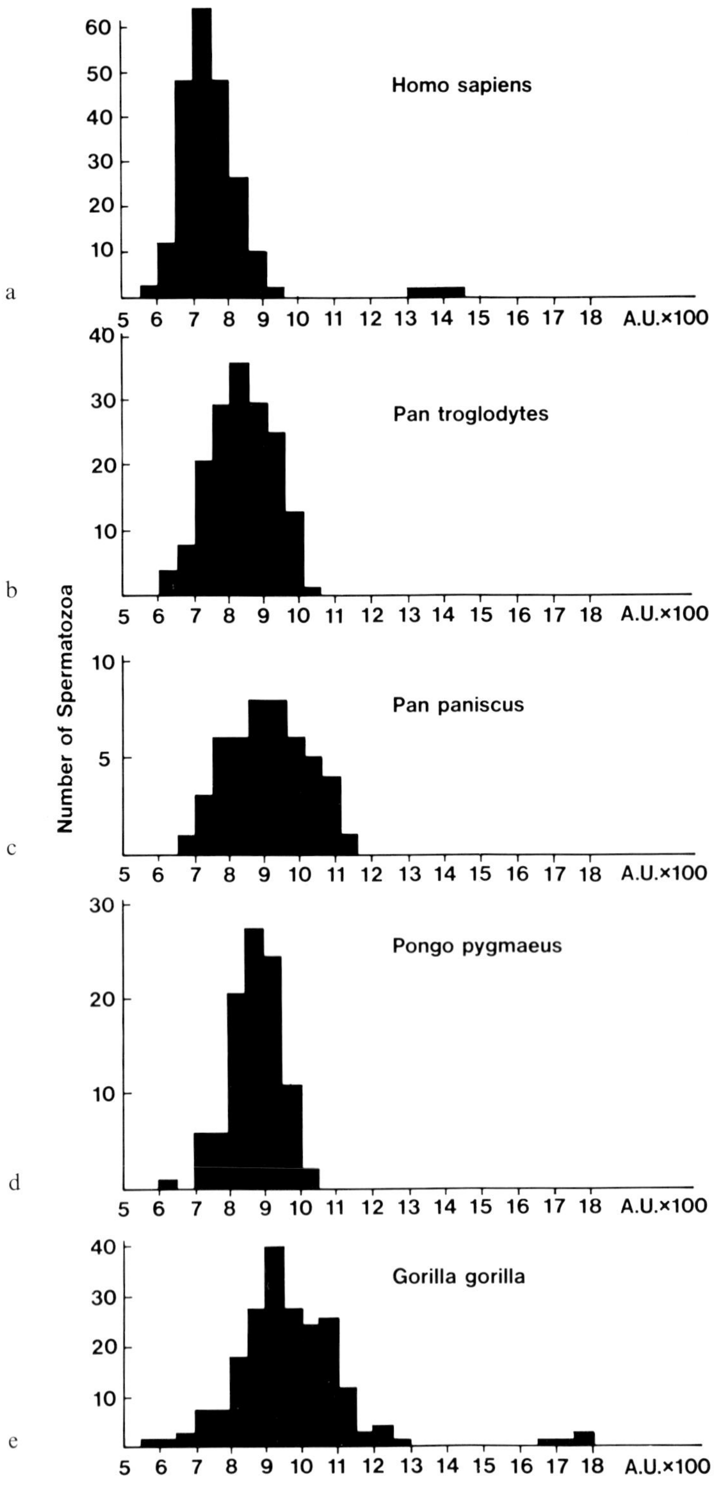
a
b
c
d
e
Number of Spermatozoa
Homo sapiens
Pan troglodytes
Pan paniscus
Pongo pygmaeus
Gorilla gorilla
A.U.×100
5 6 7 8 9 10 11 12 13 14 15 16 17 18
10 20 30 40 50 60
5 10

Table 18.1. Comparisons of total dry mass of spermatozoa of the great apes and man

Species	Means	Difference	Approximate 95% confidence limits			
			Assumption A		Assumption B	
			Lower	Upper	Lower	Upper
Chimpanzee–Human	810.0–742.3	+ 67.7	+ 21.8	+113.6	+ 41.8	+ 93.6
Orangutan–Human	871.6–742.3	+129.3	+ 73.0	+185.6	+ 97.6	+161.0
Gorilla–Human	965.2–742.3	+222.9	+166.4	+279.5	+190.7	+255.2
Orangutan–Chimpanzee	871.6–810.0	+ 61.6	+ 5.3	+117.9	+ 29.9	+ 93.3
Gorilla–Chimpanzee	965.2–810.0	+155.2	+ 98.7	+211.8	+123.0	+187.5
Gorilla–Orangutan	965.2–871.6	+ 93.6	+ 28.5	+158.9	+ 56.6	+130.7

The component of variance between individuals within a species is taken to be either 2000 a.u.^{2} (assumption A) or 1000 a.u.^{2} (assumption B). Analysis of variance of haploid dry-mass revealed (1) no significant differences between preparations from the same individual, (2) that the intra-individual variance components for man, chimpanzee, and orangutan did not differ significantly (pooled estimate = 4949.3 with 485 d.f.), but were significantly less than for gorilla (11 606.7 with 197 d.f.). Confidence intervals were constructed using these estimates and taking values of the inter-individual intra-species component to be either 2000 a.u.^{2} (assumption A) or 1000 a.u.^{2} (assumption B). The assumptions correspond to a situation in which one individual in 100 can be expected to differ from its species mean by the equivalent of approximately two (A) or one (B) medium-sized chromosomes respectively. The implied degree of aneuploidy suggests that these are overestimates, as are the widths of the corresponding confidence intervals. [Seuánez et al.: Nature (London) *270*, 345–347 (1977)].

are therefore allowed to degenerate or disappear. There seems to be no correlation between DNA content and chromosome numbers in mammals (Bachmann, 1972), and the fact that man has 46 chromosomes against 48 of the great apes, as a result of a telomeric fusion, cannot explain the differences of DNA content in different species. Manfredi-Romanini and Campana (1971) compared the DNA content of two populations of rats, one of which had suffered a Robertsonian fusion. No significant difference was detected between them as a result of the chromosome rearrangement. Though not identical, this kind of rearrangement is comparable to the one that took place in the evolutionary branch leading to man, accounting for the formation of chromosome 2 (HSA 2).

Fig. 18.2. The estimations of total dry mass (TDM) were obtained with a Vickers M-86 integrating microinterferometer. This instrument measures the optical path difference (o.p.d. = refractive index × thickness) in arbitrary units (a.u.). The area of the object (the sperm head only) was selected using an electronic masking system. Measurements were obtained with a ×75 n.a. 1.1 water immersion objective and preparations were measured while immersed in distilled water. Fifty cells were measured per individual using two slides (25 measurements in each), except in the gorillas (100 cells; 50 measurements per slide). a *Homo sapiens*; b *Pan troglodytes*; c *Pan paniscus*; d *Pongo pygmaeus*; e *Gorilla gorilla*. [Seuanez et al., Nature (London) *270*, 345–347 (1977)]

Moreover, the DNA content of two phylogenetically related species with equal chromosome numbers and very similar banding patterns, such as the ox and the goat (Evans et al., 1973), have been shown to have differences in their DNA content of approximately 15% (Sumner and Buckland, 1976). Since the amount of satellite DNA is approximately the same in these two species, it was suggested that the extra DNA of the ox must be evenly distributed throughout the genome. In the Hominidae, where chromosome numbers are 48 in four species and 46 in one species, and 99% of the G-(or R-)chromosome bands are common to all species (Dutrillaux, 1975), this might also be the case. There is evidence that the great apes might have a larger amount of highly repetitive DNA sequences than man as indicated by the number of grains in autoradiographic preparations after hybridisation with radioactive human cRNAs (Chap. 14). But it is improbable that such differences in highly repetitive DNAs between these species could explain differences between total amounts of DNA, since in man, the amount of satellite DNA is very low, and probably represents no more than 6% of the diploid genome. It is probable that in the great apes, as in man, satellite DNA may comprise a similar proportion of the total genome, as was suggested by a comparison of Cot curves of DNA reassociation (Chap. 13). Thus, the reason why man has less DNA than any of his closest living relatives is still obscure, and no simple straightforward explanation can be advanced at present.

As a final conclusion to this chapter and to this book I feel it is necessary to emphasise how far we still are from having a clear idea of how our own species has evolved. With the development of new techniques in the years to come let us hope that substantial information will be obtained which will enlarge the limited understanding we already have of our own nature and origins. Until then, and perhaps even then, the basic question of *What is man?* will remain still unanswered.

References

Atkin, N.B., Mattison, G., Beçak, W., Ohno, S.: The comparative DNA content of 19 species of placental mammals, reptiles and birds. Chromosoma (Berlin) *17*, 1–10 (1965)

Bachmann, K.: Genome size in mammals. Chromosome (Berlin) *37*, 85–93 (1972)

Bedi, K.S., Goldstein, D.J.: Cytophotometric factors causing apparent differences between Feulgen DNA contents of different leukocyte types. Nature (London) *251*, 439–440 (1974)

Dutrillaux, B.: Sur le nature el l'origine des chromosomes humaines. Paris: L'expansion Scientifique 1975

Evans, H.J., Buckland, R.A., Sumner, A.T.: Chromosome homology and heterochromatin in goat, sheep and ox studied by banding techniques. Chromosoma (Berlin) 42, 383–402 (1973)

Hinegardner, R.: Evolution of genome size. In: Molecular evolution. Ayala, F. (ed.), pp. 179–199. Sunderland, Massachusetts. Sinauer Associated Inc. Publishers 1976

Hotta, Y., Chandley, A.C., Stern, H.: Biochemical analysis of meiosis in the male mouse. Chromosoma (Berlin) 62, 255–268 (1977)

Hotta, Y., Ito, M., Stern, H.: Synthesis of DNA during meiosis. Proc. Nat. Acad. Sci. USA 56, 1184–1191 (1966)

Lima de Faria, A., German, J., Ghatnekar, M., McGovern, J., Anderson, L.: In vitro labelling of human meiotic chromosomes with H^3 thymidine. Hereditas 60, 249–261 (1968)

Manfredi-Romanini, M.G.: Nuclear DNA content and area of primate limphocytes as a cytotaxonomical tool. J. Hum. Evol. 1, 23–40 (1972)

Manfredi-Romanini, M.G.: The DNA nuclear content and the evolution of vertebrates. In: Cytotaxonomy and vertebrate evolution. Chiarelli, B., Campana, E. (eds.), pp. 39–81. New York, London: Academic Press 1973

Manfredi-Romanini, M.G., Campana, E.: Contenuto in DNA nei nuclei postcinetici in due popolazioni cromosomicamente differenti di *Rattus ratus,* cited by Manfredi-Romanini, M.G. The DNA nuclear content and the evolution of vertebrates. In: Cytotaxonomy and vertebrate evolution. Chiarelli, B., Campana, E. (eds.), pp. 39–81. New York, London: Academic Press 1973

Mirsky, A.E., Ris, H.: The deoxyribonucleic acid content of animal cells and its evolutionary significance. J. Genet. Physiol. 34, 451–462 (1951)

Monesi, V.: Chromosome activities during meiosis and spermiogenesis. J. Reprod. Fertil. Suppl. 13, 1–14 (1971)

Seuánez, H.N., Carothers, A.D., Martin, D.E., Short, R.V.: Morphological abnormalities in the spermatozoa of man and the great apes. Nature (London) 270, 345–347 (1977)

Sparrow, A.H., Price, H.J., Underbrink, A.G.: A survey of DNA content per cell and per chromosome of prokaryotic and eukaryotic organisms: some evolutionary considerations. In: Evolution of genetic systems. Smith, H.H. et al. (ed.) Brookhaven Symp. Biol. 23, 451–495 (1972)

Sumner, A.T., Buckland, R.A.: Relative DNA contents of somatic nuclei of ox, sheep and goat. Chromosoma (Berlin) 57, 171–175 (1976)

Taylor, H., McMaster, R.D.: Autoradiographic and microphotometric studies of DNA during microgametogenesis in *Lilium longifluorum.* Chromosoma (Berlin) 6, 489–521 (1954)

Vendrely, R., Vendrely, C.: La teneur du noyau cellulaire en ADN à travers les organes, les individus et les espèces animals. Experientia 4, 434 (1948)

Vendrely, R., Vendrely, C.: L'acide désoxyribonucleique. Substance fondamentale de la cellule vivante. Paris: A. Legrand et C. 1957

Subject Index

Allopatric speciation 79–82
Amino Acid substitutions 15, 16, 101–107 (see also mutation)
Ammospermophilus leucurus 134
Ancestral karyotype of the Hominidae 71–77
Aotus trivirgatus 160
Apomorph characteristics 72, 73, 75, 76
Ateles geoffrey 160
Australopithecus 8–11, 13
  africanus 4, 8–10
  boisei 9, 81 (see *Zinjanthropus boisei*)
  robustus 9, 10 (see *Paranthropus robustus*)

Baleanoptera borealis 109

Carollia perspicillata 173
Cellular heterokaryons 111
Centric fission 70, 119, 120
Centric fusion 23, 117–120, 183
Cercocebus
  atys 98
  galeritus 179
  torquatus 179
Cercopithecus aethiops (African green monkey) 70, 120, 128, 129, 132
Cervus
  canadiensis 79
  elephus 79
Chromosomal theory 23
Chromosome evolution 79–82 (see rates)
  in Cetacea 80, 81, 109
  in Hominidae 65–77
  in Hominoidea-Cercopithecoidea 115–120
  in Pinnipedia 80, 81
  in Rodentia 80, 81
Colobus pokylomus 160
Comparative gene mapping of single gene sequences 42, 70, 111, 121, 122, 145 (see Syntenic groups in primates)
Comparative gene mapping of repetitive sequences – see Ribosomal DNA sequences and Satellite DNA
Cricetulus griseus 174

Delphinus dolphin 109
DNA content, evolutionary implications 181, 183, 184 (see polyploidy)
  of bacteria 179
  of Ceboid monkeys 179
  of Cercopithecoid monkeys 179
  of gibbons 128, 179
  of goat 184
  of great apes 128, 179
  of man 128, 179
  of ox 184
  of placental mammals 179
  of salamanders 179
  of spermatozoa of man and great apes 179–183
DNA heteroduplex molecules 95–99, 130, 131
DNA homoduplex molecules 95–99, 130, 131
DNA reassociation analysis 87–89, 128, 130
DNA, repetitive fractions 16, 24, 37, 87–93, 106, 128, 184 (see Palindromes, Ribosomal DNA sequences, and Satellite DNA)
  in African green monkey 128, 129
  in Cercopithecoidea 128–130
  in Hominoidea 128, 130, 131
  in Lorisoidea 130
  in man 128, 130, 184
  in *Pan troglodytes* 128–130
  in Primates 128
  in Rhesus monkey 128, 130
DNA, replication patterns in chromosomes 164–176
  in phylogenetically related species 175, 176 (see *Mus* and *Dipodomys*)
  in relation to banding in Hominidae 165–172
  in relation to euchromatin and heterochromatin 172–174
  in the X chromosome of the Hominidae 170–172
DNA, thermal stability analysis 91–93, 129–131
DNA, type C-viral sequence 98, 99
DNA, unique copy (non-repetitive) 87, 88, 95–99
Dipodomys 134
  merriani 173, 174
  ordii 134
  panamintinus 173
Drosophila 91
  melanogaster 147

Dryopithecus 11, 12
   africanus 12
   major 12

Elephe radiata 133
Equus 15
Eukaryote genome (composition) 87–89, 106

Fitness 21
   multiplicative 21, 22, 105, 106
   threshold 21, 22, 105, 106
Frozen accidents 120–125

Gallus domesticus 133
Gene duplication 22, 122–125 (see Polyploidy)
Gene fixation 21, 22, 106 (see Rates)
Generation time
   of bacteria 3
   of man 3
   of mouse 3
Genetic load 16
Germ cell divisions
   in human female 4
   in human lineage 4
   in human male 4
   in male mouse 4
Gigantopithecus 12
Gorilla gorilla 8, 12, 17, 18, 27
Gorilla gorilla (chromosomes)
   Ammoniacal silver staining 36, 157–159 (see Ribosomal DNA sequences)
   C-banding 32–34, 53, 54, 147–152
   G-11 staining 34, 35, 153–155
   G- (or R-) banding 28, 31, 151
   Heteromorphisms 51, 53, 54
   Homologies with human and ape chromosomes 36–42 (see Syntenic groups in Primates)
   Late DNA replicating patterns 165–172
   Methylated DNA sequences 35, 76, 77
   Normal diploid number 18, 27
   Q-banding 29–32, 34, 53, 54, 141, 149, 151, 152
   T-banding 36

Heterochromatin
   constitutive 172, 173
   definition 172, 173
   DNA composition in Hominidae (see Satellite DNA)
   DNA composition in man (see Satellite DNA)
   facultative 172
   replication patterns 172–174
Hominoidea- (classification) 17–19
Homo 10
   erectus 10–13, 81 (see *Pithecanthropus erectus*)
   habilis 10, 13, 81
   origin of the genus 10
   sapiens 3, 13, 17, 18, 82, 111
   sapiens neanderthalensis 13
   sapiens sapiens 13
Human chromosomes
   Ammoniacal silver staining 36, 157–159 (see Ribosomal DNA sequences)
   C-banding 32, 33, 47–50, 147–152
   G-11 staining 34, 153–155
   G- (or R-) banding 28, 33, 151
   Heteromorphisms 46–50
   Homologies with ape chromosomes 36–42 (see Syntenic groups in Primates)
   Late DNA replicating patterns 164–166, 170–173
   Methylated DNA sequences 35, 76, 77
   Normal diploid number 18, 27
   Numerical aberrations 46, 47
   Polymorphisms 47–50 (see Y chromosome)
   Q-banding 29–31, 34, 47, 48, 138, 151
   T-banding 35, 36
Human distinctiveness 5, 6
Human origin
   African 8, 9, 12
   Asian 8–12, 98, 99
   Monocentric 13
   Polycentric 13
   Single species hypothesis 82
Hylobates 17, 18, 99
   concolor 17, 42
   lar 17, 42, 70, 72, 160
   moloch 17, 69, 70, 72

In situ hybridisation 24, 34, 68, 71, 131, 133, 137, 138, 140–155, 157, 159–161
Interferometry 180–182
Intraspecific variation 20
Inversions 23, 47–49, 65, 69, 111, 123 (see Paracentric, and Pericentric inversions)

Lemur fulvis 160

Macacca mulatta 98, 113–116, 160
Meganthropus paleojavanicus 10, 11
Microdensitometry 49, 50, 179, 180
Microtus
   agrestis 152
   oeconomus 70
Molecular evolution 15, 16 (see Amino Acid substitutions, Mutation, Nucleotide substitutions, and Protein Evolution)
   in relation to the classification of primates 17, 18, 106
Molecular evolutionary clocks 104–106
Muntiacus 109
   muntjack 109, 174
   reevesi 109

Mus 175
  dunni 175
  fulvidiventris 175
  musculus 173, 175
  musculus mollosinus 175
Mutation (see Rates)
  in baboon type C-viral DNA 98, 99
  back mutation 104, 105
  frameshift 101
  in germinal cell lines 3, 4
  mechanisms 3
  missense 101
  neutral 105
  nonsense 101
  in regulatory genes 108
  samesense 101
  in structural genes 5, 6, 16, 80, 101–108
  in tRNA 101

Nasalis nervatus 179
Natural Selection 4, 20–22, 80, 105
  cost of selection 21, 22, 105, 106
Neutral mutation theory 104–106
Nucleotide substitutions 16, 17, 95–99, 101
  (see Rates)

Organic evolution 15, 17, 108, 109 (see Rates)
  and the classification of primates 17–19
Palindromes 88 (see DNA, repetitive)
Pan paniscus 17, 18, 27
Pan paniscus (chromosomes)
  Ammoniacal silver staining 36, 157–159 (see Ribosomal DNA sequences)
  C-banding 32–34, 51–53
  G-11 staining 53
  G- (or R-) banding 28, 30, 69
  Heteromorphisms 51–53
  Homologies with human and ape chromosomes 36–42 (see Syntenic groups in Primates)
  Normal diploid number 18, 27
  Q-banding 29–32, 52, 53
  T-banding 36
Pan troglodytes 8, 12, 17, 18, 27
Pan troglodytes (chromosomes)
  Ammoniacal silver staining 36, 157–159 (see Ribosomal DNA sequences)
  C-banding 32–34, 51–53, 147–152
  G-11 staining 34, 35, 153–155
  G- (or R-) banding 28, 29, 69, 151
  Heteromorphisms 51, 52
  Homologies with human and ape chromosomes 36–42 (see Syntenic groups in Primates)
  Methylated DNA sequences 35, 76, 77
  Normal diploid number 18, 27
  Q-banding 29–32, 34, 51–53, 148, 151, 152

T-banding 36
Trisomy 27
Papio
  cynocephalus 98, 116, 118, 128, 160, 161
  hamadryas 160
  papio 70, 116–120
Paranthropus robustus 9 (see *Australopithecus robustus*)
Paracentric inversions 41, 65, 67, 74, 75, 117–120
Pericentric inversions 39, 41, 65–67, 74, 75, 80
Parsimonious evolution 15, 41, 106, 107
Phyletic relationship of man 17–19, 75–77, 97–99, 104–106
Phylogenetic trees
  additive 102–106
  divergence 71, 72, 102
  maximum parsimony 15, 106, 107
Pithecanthropus
  erectus 8, 9 (see *Homo erectus*)
  modjokertensis 11, 12
Pithecia pithecia 160
Plesiomorph characteristics 72–75
Polyploidy 122–125
Pongo pygmaeus 8, 17, 18, 27, 99
Pongo pygmaeus (chromosomes)
  Ammoniacal silver staining 36, 157–159 (see Ribosomal DNA sequences)
  C-banding 32, 33, 147–152
  Difference between Bornean and Sumatran animals 55, 56, 58, 59, 66, 81
  Double inversion polymorphism 56, 57, 60–62, 66, 80, 81
  G-11 staining 35, 153–155
  G- (or R-) banding 28, 32, 55–59
  Heteromorphisms 51, 55–62
  Homologies with human and ape chromosomes 36–42 (see Syntenic groups in Primates)
  Late DNA replicating patterns 165–172
  Normal diploid number 18, 27
  Q-banding 29–32, 55, 56, 150, 152
  T-banding 36
Presbytis entellus 70
Protein evolution 6, 15, 16, 101–108 (see Amino Acid substitutions, Molecular evolution, Mutation, and Rates)
Ptyas mucosas 133, 134

Ramapithecus 11–13, 105
  sensu strictu 11
Rana 108
  catesbiana 108
  esculenta 108
Rates
  of amino acid substitutions 105–108
  of chromosome evolution in frogs 108, 109

of chromosome evolution in Hominidae 75
of chromosome evolution in mammals 108, 109
of fixation of mutations   4, 21, 22, 104–108 (see Gene fixation)
of lethal and deleterious mutations   16 (see Genetic load)
of nucleotide substitutions in primates 95–97
of organic evolution   15, 24, 108, 109
Rattus   69
  rattus   61
Reproductive isolation   22, 23, 79, 80, 82
Ribosomal DNA sequences   157
  5S sequences in man   88, 145, 157, 161
  5S sequences in great apes   145, 161
  5S sequences in *Papio cynocephalus*   118, 161
  18S and 28S sequences in man   88, 157–159
  18S and 28S sequences in great apes   38, 39, 68, 158, 159, 161
  18S and 28S sequences in other primates 159–161

Saimiri sciureus   69, 160
Sanguinus nigricollis   160
Satellite DNA   89 (see DNA, repetitive, and In situ hybridisation)
  in African green monkey   129, 132
  in *Ammospermophilus leucurus*   134
  in baboon   129
  in *Dipodomys ordii*   134
  in *Drosophila*   91
  in *Elephe radiata*   133
  evolutionary considerations in general 131–135
  female specific in reptiles and birds   133, 134
  in Guinea pig   91, 134
  in man   18, 19, 89–93, 130–132, 184
  in man, base composition of satellites   91
  in man, distribution in human chromosomes   50, 137–139
  in man, distribution of homologous sequences in ape chromosomes   34, 39, 68, 71, 139–146
  in man, evolutionary considerations   93, 130–133, 145, 146
  in man, nomenclature and physical properties of satellite DNA fractions 90–93
  in man, relation between satellite III DNA and G-11 staining in the Hominidae 35, 153–155
  in man, restriction pattern of satellite III and male specific DNA   131, 132
  in man, in relation to C-banding in the Hominidae   147–153
  in man, thermal denaturation analysis of satellite DNA   91–93, 132
  in *Pan troglodytes*   129, 131
  restriction pattern in different mammals 132
  in *Thonomis bottae*   134
Sivapithecus   12
Stasipatric specistion   79–82
Symphalangus   18, 99
  syndactylus   17, 160
Syntenic groups in Primates   18, 24, 65, 111–122

Tandem repeats (of a DNA sequence)   88 (see DNA, repetitive)
Telomeric fusion   41, 65, 67, 68, 70, 74, 111, 117, 118, 183
Thalarctos maritimus   79
Thonomis bottae   134
Translocations   60, 70, 111, 183
Tupaia tupaia   160

Ursus actos   79

X chromosome
  in the Chinese hamster   174
  in Gorilla   31, 40, 41, 54, 154, 155, 170–172
  in the Hominidae   36, 38–41, 72–74, 114, 170–172
  in Hylobates   72
  in mammals   120, 121, 175, 176
  in man   33, 40, 41, 164, 166, 170–172
  in *Pan paniscus*   30, 40, 41
  in *Pan troglodytes*   29, 40, 41, 170–172
  in *Pongo pygmaeus*   31, 40, 41, 170–172
  in Primates   116
Xenopus   157
"Xenopus pattern"   88

Y chromosome
  in *Cercopithecus diana*   42
  in the Chinese hamster   174
  in Gorilla   31, 35, 42, 53, 54, 76, 77, 141, 143, 148, 149, 152–154
  in Hominidae   42, 73
  in *Hylobates concolor* and *lar*   42
  in man   31, 35, 42, 47–50, 76, 77, 132, 137, 138, 142, 143, 147
  in mouse   147
  in *Pan paniscus*   30, 31
  in *Pan troglodytes*   31, 42, 142, 143, 148, 149, 152–154
  in *Pongo pygmaeus*   32, 142, 143, 148, 150, 152, 154
  in Primates   42

Zinjanthropus boisei   9 (see *Australopithecus boisei*)

CHROMOSOMA

ISSN 0009-5915   Title No. 412

Editorial Board: H. Bauer (Managing Editor), Erlangen; W. Beermann, Tübingen; J. G. Gall, New Haven, CT; B. John, Canberra; H. C. Macgregor, Leicester; R. B. Nicklas, Durham; M. L. Pardue, Cambridge, MA; J. Sybenga, Wageningen; J. H. Taylor, Tallahassee, FL; D. von Wettstein, Copenhagen

Advisory Board: R. Dietz, Tübingen; J. E. Edström, Stockholm; W. Hennig, Nijmegen; T. C. Hsu, Houston; H.-G. Keyl, Bochum; C. D. Laird, Seattle, WA; G. F. Meyer, Tübingen; D. Schweizer, Wien; H. Stern, La Jolla, CA.

CHROMOSOMA, founded in 1939, publishes original contributions concerning all aspects of nuclear and chromosome research. Current studies in this field range from those on protozoan chromosomes to research on the nuclei of higher organisms and frequently apply techniques and material from fields as diverse as molecular and population genetics. Pertinent biochemical and biophysical approaches to cytological problems are often included.

Fields of Interest: Genetics, Cytology, Molecular Biology, Biochemistry, Virology, Microbiology, Botany, Zoology, Biology, Anatomy, Developmental Physiology.

Subscription Information and sample copies upon request.

Springer-Verlag
Berlin
Heidelberg
New York

Human Genetics

ISSN 0340-6717   Title No. 439

Editorial Board: W. Lenz, Münster; A. G. Motulsky, Seattle; F. Vogel, Heidelberg; U. Wolf, Freiburg i. Br.;

Advisory Board: G. Anders, Groningen; H. Baitsch, Ulm; A. G. Bearn, New York; H. Bickel, Heidelberg; N. P. Bochkov, Moscow; D. Bootsma, Rotterdam; K. H. Degenhardt, Frankfurt/M.; B. Dutrillaux, Paris; G. Flatz, Hannover; U. Francke, New Haven; W. Fuhrmann, Giessen; H. Grüneberg, London; B. Hassenstein, Freiburg i. Br.; K. Hirschhorn, New York; P. S. Jacobs, Honolulu; W. Jaeger, Heidelberg; D. Klein, Genève; E. Krah, Heidelberg; W. Krone, Ulm; H. Lehmann, Cambridge; V. A. McKusick, Baltimore; M. Mikkelsen, Glostrup; O. J. Miller, New York; H. Nachtsheim, Boppard; E. Passarge, Essen; A. Prader, Zurich; H. Ritter, Tübingen; D. F. Roberts, New Castle/T.; C. Ropartz, Bois-Guillaume; W. Schmid, Zurich; U. W. Schnyder, Heidelberg; W. J. Schull, Houston; H. G. Schwarzacher, Wien; C. Stern, Berkeley; H. E. Sutton, Austin

Recent years have witnessed considerable progress in general and human genetics. New concepts about the structure and function of the gene are of major importance for the understanding of human genetics. Simultaneously, methodological advances have led to new insights concerning the genetic basis of health and disease in man. Research in cytogenetics, biochemical genetics, population genetics, immunogenetics, and pharmacogenetics now strongly supplements studies in formal genetics. Based on established scientific traditions, this journal reports and summarizes new developments in human and medical genetics.

For subscription information and for your free sample copy please write to:

Springer-Verlag, Wissenschaftliche Information, Postfach 105280, D-6900 Heidelberg, FRG

R. Rieger, A. Michaelis, M. M. Green
Glossary of Genetics and Cytogenetics
Classical and Molecular
Springer Study Edition
4th completely revised edition. 1976.
100 figures, 8 tables. 647 pages
ISBN 3-540-07668-9

The fourth edition of the wellproven Glossary is now complete, and has been completely reworked and enlarged since the third edition. Special care has been given to including the important new terms which have meanwhile come to be used in genetics. The style of the Glossary has been preserved, so that only a definition of scientific expressions is given when that suffices, but when a term can only be understood by more detailed explanation, it is described in a short essay, including scientific data. The inclusion of a short history of the subject, together with the crossreferences and quotations from the literature, make this Glossary a handbook for scientists and a valuable textbook for students. The lower price of this edition as compared to the last should encourage a wide distribution of this new work.

P. T. Kelly
Dealing with Dilemma
A Manual for Genetic Counselors
1977. 1 figure. XIII, 143 pages
(Heidelberg Science Library)
ISBN 3-540-90237-6

Contents:
Introduction. – Overview of the Genetic Counseling Process. – Intake Visit. – Diagnostic Visit. – Follow-Up Visits. – Emotional and Social Reactions to Genetic Disease. – Genetic Counseling Techniques. – Sociological Aspects of Genetic Counseling. – Questions Genetic Counselors Ask. – Appendices.

This manual has been written to fill a gap that exists within the field of Genetic Counseling. The few genetic counseling books currently available are devoted almost exclusively to medical and genetic aspects. No book yet exists to provide genetic counselors with guidance and information about the social-psychological aspects of genetic counseling – those aspects which enable them to be truly counselors, not merely transmitters of scientific or medical information. "Dealing With Dilemma" contains numerous examples taken from actual genetic counseling situations (based on the author's experience as a genetic counselor and as a teacher of genetic counseling), as well as ways to deal with specific genetic counseling problems. This book is written in nontechnical language. It presupposes some knowledge of Mendelian genetics, polygenic inheritance, and chromosomal anomalies, but no prior study of psychology or counseling techniques. In comprehensive fashion, it integrates medical and genetic information with the psychological aspects of genetic counseling to provide a working manual for genetic counselors.

Springer-Verlag
Berlin
Heidelberg
New York